Practical Statistics
for Educators

Practical Statistics for Educators

Fifth Edition

RUTH RAVID

ROWMAN & LITTLEFIELD
Lanham • Boulder • New York • London

Published by Rowman & Littlefield
A wholly owned subsidiary of The Rowman & Littlefield Publishing Group, Inc.
4501 Forbes Boulevard, Suite 200, Lanham, Maryland 20706
www.rowman.com

Unit A, Whitacre Mews, 24-26 Stannary Street, London SE 11 4AB

Copyright © 2015 by Rowman & Littlefield

British Library Cataloguing in Publication Information Available

Library of Congress Cataloging-in-Publication Data
Ravid, Ruth.
  Practical statistics for educators / Ruth Ravid. — Fifth edition
      pages cm
  Includes bibliographical references and index.
  ISBN 978-1-4422-4285-2 (cloth) — ISBN 978-1-4422-4286-9 (pbk.)
  — ISBN 978-1-4422-4287-6 (electronic) 1. Educational statistics--Study and teaching. 2. Educational tests and measurements. I. Title.
  LB2846.R33 2015
  370.2'1—dc23
                                                              2014030157

Contents

List of Statistical Symbols   vii

Preface   ix

PART I: Introduction

 1   An Overview of Educational Research   3

 2   Basic Concepts in Statistics   21

PART II: Describing Distributions

 3   Organizing and Graphing Data   49

 4   Measures of Central Tendency   69

 5   Measures of Variability   77

 6   The Normal Curve and Standard Scores   89

PART III: Measuring Relationships

 7   Correlation   103

 8   Prediction and Regression   119

PART IV: Comparing Group Means

 9   t Test   133

10   Analysis of Variance   149

PART V: Chi Square Test

11   Chi Square Test   173

PART VI: Standardized Test Scores, Reliability, and Validity

12 Interpreting Standardized Test Scores   185

13 Reliability   193

14 Validity   205

PART VII: Putting It All Together

15 Choosing the Right Statistical Test   215

16 Planning and Conducting Research Studies   225

17 Using Statistical Tests to Analyze Survey Data   239

Glossary   253

Index   265

About the Author   275

List of Statistical Symbols

H_A Alternative (research) hypothesis; also represented by H_1 (Ch. 2)

H_O Null hypothesis (Ch. 2)

p Probability; level of significance (Ch. 2)

α Probability level set at the beginning of the study (Ch. 2)

df Degrees of freedom (Ch. 2)

ES Effect size (Ch. 2)

r Pearson's correlation coefficient (also an index of reliability) (Ch. 2, 7)

r^2 Effect size; coefficient of determination (Ch. 2, 7)

$SE_{\bar{x}}$ Standard error of the mean (Ch. 2)

CI Confidence interval (Ch. 2)

X Raw score (Ch. 4)

N Number of people in a group (or in a population) (Ch. 4)

n Number of people in a group (or in a sample) (Ch. 4)

Σ Sum of (Greek letter sigma, uppercase) (Ch. 4)

$\bar{x}$ Mean of sample (Ch. 4)

μ Mean of population (Greek letter *mu*) (Ch. 4)

S Standard deviation (SD) of sample (Ch. 5)

S^2 Variance of sample (Ch. 5)

σ Standard deviation (SD) of population (Greek letter *sigma*, lowercase) (Ch. 5)

σ^2 Variance of population (Ch. 5)

z z score (Ch. 6)

T T score (Ch. 6)

Y' Predicted Y score (in regression) (Ch. 8)

b Slope (or *coefficient*; in regression) (Ch. 8)

a Intercept (or *constant*; in regression) (Ch. 8)

S_E Standard error of estimate (in regression) (Ch. 8)

R Multiple correlation coefficient (Ch. 8)

R^2 Coefficient of determination of multiple correlation (in regression) (Ch. 8)

t t value (Ch. 9)

F F ratio (Ch. 10)

K Number of groups (in ANOVA) (Ch. 10)

SS   Sum of squares (in ANOVA) (Ch. 10)
MS   Mean squares (in ANOVA) (Ch. 10)
χ^2 Chi square value (Ch. 11)
SEM   Standard error of measurement (Ch. 13)

Preface

There are many introductory statistics textbooks but only a few of them focus on the practical use of statistics in education. Today, more than ever before, educators are required to gather and analyze various forms of quantitative data for the purpose of assessment and for data-driven decision making. *Practical Statistics for Educators* is a clear and easy-to-follow text written specifically for education students in introductory statistics and action research courses. It is also an invaluable resource and guidebook for educational practitioners who wish to study their own settings.

The book introduces educational students and practitioners to the use of statistics in education; basic concepts in statistics are explained in clear language. All of the examples in the book that demonstrate the use of statistics in research are taken from the field of education and serve to illustrate the various concepts, terms, statistical tests, and data interpretations discussed in the book. Formulas and equations are used sparingly in order to explain certain points and to show the computations of statistical tests. Therefore, readers are not required to do any computations and can simply follow the steps that are presented.

The topics of testing, test score interpretation, reliability, and validity are included to help educators understand these essential topics. These topics are often missing from basic statistics books but are critical to educators' understanding of assessment in the K–12 system.

Chapter previews provide a quick glance of the information provided in the chapter. Succinct summaries highlighting each chapter's main points are presented at the end for a quick review. A glossary of main terms and concepts helps readers navigate the book and easily find the term they are looking for, including the chapter where the term is introduced. The detailed index is another excellent tool for finding terms and the place in the text where they are discussed.

The focus of the book is on essential concepts in educational statistics, understanding when to use various statistical tests, and how to interpret results. For the practitioner-researcher there is information about planning the study and reporting the results. The book also helps readers become more knowledgeable researchers by better understanding and being more informed consumers of published research.

The main changes to the fifth edition are as follows:

- A new chapter was added on the topic of analyzing survey data. The chapter starts with an introduction to survey design and construction and is followed by examples of the use of statistical tests to analyze survey data. (The statistical tests include Pearson correlation, *t* test, ANOVA, and chi square.) The additional chapter is in part VII, which is newly titled *Putting It All Together*. The three chapters in this part allow readers to consider the information presented in the book in the context of their own research or to analyze data collected in their own classrooms, programs, or class.
- Exercises and activities were added at the end of the chapters to encourage readers to apply the information presented in the chapter to their own settings.

Additional changes were made in the reorganization of text throughout the book to facilitate a better flow. These changes include: (a) moving the chapter titled "The Normal Curve and Standard Scores" to part II, which is now titled *Describing Distributions*; (b) moving the chapter titled "Interpreting Standardized Test Scores" to part VI just before "Reliability" and "Validity"; (c) moving the chapter on chi square to its own part (part V); and (d) moving the discussion of skewed distributions from chapter 4 to chapter 6. Further, several graphics were updated or revised, and the text has been revised and updated throughout.

Readers of this new edition will find the *Study Guide to Accompany Practical Statistics for Educators* a very helpful tool. The *Guide* offers readers the opportunity to review, practice, and apply what they have just learned.

ACKNOWLEDGMENTS

Thanks go to the team at Rowman & Littlefield: my editor, Susanne Canavan, who patiently guided the book through the revision process; Andrea O. Kendrick, assistant acquisitions editor, whose advice and prompt and clear instructions made the process go smoothly; and to production editor Laura Reiter. Special thanks go to Donna Rafanello, whose insightful edits and suggestions were invaluable in writing this edition. Thanks also to Amy Charlson, who created several new graphics and revised others.

Mostly, my thanks go to my family for their constant love and support. Thank you Cory, Ossie, Jackie, Lindsey, Ashley, and Audrey for enriching my life!

I

INTRODUCTION

1

An Overview of Educational Research

In chapter 1, you are going to learn about various approaches to research in education and some ways researchers gather information effectively. We go over three different research approaches (*basic*, *applied*, and *action* research) and learn about their advantages and limitations. Within each approach, researchers can decide which type of research to use—*qualitative* or *quantitative*. Differences between the two types are presented and explained. Researchers also have to choose between *experimental* and *nonexperimental* designs for their studies. Experimental research is discussed in greater detail, and explanations of *threats to internal and external validity* and *groups and individual designs* are included. The discussion of nonexperimental research includes explanations of *causal comparative* and *descriptive* research.

By the end of this chapter, you will know the differences among all of these terms and the advantages and disadvantages of using one approach over another. As a consumer and reader of research, you will understand the approaches and designs used by researchers. As a producer of research, this chapter will enhance your ability to plan and carry out your own investigation.

For many people, the term *research* conjures up a picture of a lab, researchers working at computers and "crunching numbers," or mice being injected with experimental drugs. Clearly, this is not what we mean by this term. In this book, we define *research* as a systematic inquiry that includes data collection and analysis. The goal of research is to describe, explain, or predict present or future phenomena. In the context of education, these phenomena are most likely to be behaviors associated with the teaching and learning processes. There are several ways to classify research into categories, and each way looks at research from a different perspective. Research may be classified as: (a) *basic* (*pure*), *applied*, or *action* research; (b) *quantitative* or *qualitative* research; and (c) *experimental* or *nonexperimental* research.

BASIC (PURE), APPLIED, AND ACTION RESEARCH

Although not all textbook authors agree, most generally divide the field of research into three categories: basic (pure) research, applied research, and action research. **Basic research** is conducted mostly in labs, under tightly controlled conditions, and its main goal is to develop theories and generalities. This type of research is not aimed at solving immediate problems or at testing hypotheses. For example, scientists who worked in labs, using animals such as mice and pigeons, developed the theory of behaviorism. These early behaviorists did not have an immediate application for their theory when it was first developed.

Applied research is aimed at testing theories and applying them to specific situations. Based on previously developed theories, hypotheses are then developed and tested in studies classified as applied research. For example, based on the theory of behaviorism, educators have hypothesized that students' behaviors will be improved when tokens are used. Next, studies were conducted where tokens, such as candies, were used as an intervention to reward students whose behavior needed improvement. After the introduction of the tokens, the students' behavior was monitored and assessed to determine the effectiveness of the intervention.

Action research is conducted by practitioner-researchers in their own settings to solve a problem by studying it, proposing solutions, implementing the solutions, and assessing the effectiveness of these solutions. The process of action research is cyclical; the practitioner-researcher continues to identify a problem, propose a solution, implement the solution, and assess the outcomes. Both qualitative and quantitative data can be gathered and analyzed in action research studies.

For many years, action research has been defined as research that is conducted for the purpose of solving a local problem, without any attempt or interest in generalizing the findings beyond the immediate setting. There were those who did not view action research as serious, rigorous research. This view has changed since educators engaged in action research borrow tools from the field of applied research. For example, they have recognized the importance of the literature review and the need to examine findings from previous research. Further, educators in settings from preschool through high school who are conducting research have made great strides in sharing their findings with colleagues through a variety of means including paper presentations at conferences and in journal articles, books, monographs, and blogs, as well as through other electronic media. The end result is that although the impetus for starting practitioner research may still be a local problem, the studies themselves are much more rigorous, using tools similar to those used in applied research.

In reading educational research literature, you may come across several terms that are often used interchangeably with the term *action research*, such as the term **practitioner research**. Additional terms that may be used are *teacher research, classroom research*, and *teacher-as-researcher*. In this book, we prefer using the term *practitioner research*, because not all research and inquiry studies are undertaken for the sole purpose of bringing about an *action*. Educators also may study their practice in order to reflect, describe, predict, and compare.

Practitioner research has been embraced by educators, mostly classroom teachers, who are interested in studying their own practice without attempting to generalize their findings to other settings. These practitioners tend to use tools and procedures typical of qualitative-descriptive research, such as interviews, journals, surveys, observations, and field notes. Tools used in quantitative-experimental research are deemed by many educator-researchers as less appropriate for practitioner research because they may require sampling, random assignment of participants to groups, intervention, and manipulation of variables. However, keep in mind that practitioners also can apply many experimental approaches and collect numerical data when studying their own settings. Teachers can study individual students in their classes by using experimental designs. In studies of groups or individuals, numerical data may be collected before and after the experimental treatment in order to assess the effectiveness of the intervention.

QUANTITATIVE VERSUS QUALITATIVE RESEARCH

Most textbooks on educational research describe methods and approaches as either quantitative or qualitative. **Quantitative research** is defined in these textbooks as research that focuses on explaining cause-and-effect relationships, studies a small number of variables, and uses numerical data. Researchers conducting quantitative research usually maintain objectivity and detach themselves from the study environment. This research approach usually starts with a hypothesis, and the study is designed to test this hypothesis. Quantitative researchers believe that findings can be generalized from one setting to other similar settings and they are looking for laws, patterns, and similarities. **Qualitative research** is defined in most textbooks as that which seeks to understand social or educational phenomena. Usually in such investigations researchers focus on one or a few cases, which are studied in depth using multiple data sources. These sources are subjective in nature (for example, interviews, observations, and field notes). Qualitative research is context-based, recognizing the uniqueness of each individual and setting.

Quantitative research is *not* the same as experimental research, although a great deal of quantitative research is experimental. And, while it is true that qualitative research is descriptive, qualitative researchers also use numerical data, such as when they count events or perform certain data-reduction analyses. These quantitative and qualitative paradigms are not a simple, clear way to classify research studies because they are not two discrete sides of a coin. Rather, the paradigms are two endpoints on a continuum, and studies can be located at different points along this continuum.

Practitioner researchers tend to use the case study approach and are usually personally involved in different phases of the study. Therefore, practitioner research defies tightly controlled designs, objective data collection, random assignment of participants, sample selection from a larger population, and other characteristics that are typical of applied or basic research. Thus, many practitioner action research studies are conducted using qualitative, naturalistic paradigms. As teacher researchers, though, we should not limit ourselves to one paradigm. Our research question should guide us in determining what paradigm(s) to use for the design and implementation of our study.

In the past, researchers identified themselves as either "qualitative researchers" or "quantitative researchers," and the two paradigms were seen as completely different. Today, while recognizing the differences between the two paradigms, more and more researchers see the two as complementary and support using both in mixed-design research studies.

In a typical quantitative study, data are collected to describe phenomena or to test hypotheses. Statistical techniques are then used to analyze the data. *This book, like most other statistics textbooks, is geared toward the analysis of quantitative, numerical data.*

EXPERIMENTAL VERSUS NONEXPERIMENTAL RESEARCH

The third way to classify research is to distinguish between experimental research and nonexperimental research.[1] In **experimental research**, researchers plan an intervention and study its effect on groups or individuals. The intervention is called the **independent variable** (or *treatment*), while the outcome measure is called the **dependent variable**. The dependent variable is used to assess the effectiveness of the intervention. For example, the independent variable may be a teaching method, new curriculum, or classroom management approach. Examples of dependent variables are test scores,

---

1. HINT: Nonexperimental research may also be called *descriptive research*.

time on task, level of satisfaction, students' motivation, or choice of extracurricular activities.

Nonexperimental research may be divided into two types: causal comparative (also called *ex post facto*) and descriptive. **Causal comparative research**, like experimental research, is designed to study cause-and-effect relationships. Unlike experimental research, though, in causal comparative studies the independent variable is not manipulated for two main reasons: either it has occurred prior to the start of the study, or it is a variable that cannot be manipulated. **Descriptive research** is aimed at studying a phenomenon as it is occurring naturally, without any manipulation or intervention. Researchers are attempting to describe and study phenomena and are not investigating cause-and-effect relationships. Next we present a discussion of experimental research, followed by a brief discussion of nonexperimental research.

Experimental Research

Although most experimental studies involve the use of statistical tests to investigate and compare groups, in certain fields, such as psychology and special education, studies that focus on individuals are gaining popularity. A discussion of such studies follows a presentation of experimental group designs.

In experimental studies that are conducted to compare groups, the experimental group members receive the treatment, while members in the control group either receive the traditional approach (for example, teaching method) or do not receive any treatment. An example might be a study conducted by a high school physics teacher who wants to test the effectiveness of using an Internet website to enhance her teaching. The teacher teaches two similar-level physics classes and uses an Internet website with one of the classes (classroom 1), but not with the other (classroom 2). The Internet website used in classroom 1, which was created, moderated, and facilitated by the teacher, enables the teacher and the students to communicate with each other. The site contains materials prepared by the teacher, such as course schedule or outline, handouts, daily and weekly assignments, review exercises, suggested reading, practice tests, and other Internet sites to be used as resources,. The Internet website is also used by the students to discuss assignments, pose questions, suggest activities, and more. The students in classroom 2 serve as the control group and continue their studies using the same approaches that were used the previous year. Students in both classes are pretested and posttested on their knowledge of physics at the beginning and at the end of the semester, and gain scores are computed. The gain scores of the students in the experimental group (classroom 1) who used the class Internet website are compared to the gain scores of the students in the control group (classroom 2) to determine the effect of the Internet website on their performance in the physics class.

In other cases, no treatment is applied to control group members who are being compared to the experimental group. For example, let's say that researchers want to study the effect on preschoolers of watching a video showing young children collaborating and sharing toys. The researchers may show the video to one class of preschoolers but not to the other class. Next, all children would be observed as they interact and play with each other to determine if the children who watched the video are more likely to exhibit collaborative play behavior compared with those who did not watch the video.

Other researchers, posing the same research question about the effect of violent movies on young children, may choose another experimental design and have these children serve as their own control. They may design a study where the behaviors of the same children would be observed and studied twice: once before and once after they watch the violent movie. Then, the researchers would note any change in behavior in the children, all of whom were administered the treatment (that is, watching the movie).

As mentioned before, researchers conducting experimental research study the effect of the independent variable on the dependent variable. Still, when researchers observe changes in the dependent variable, they have to confirm that these changes have occurred as a result of the independent variable and are not due to other variables, called extraneous variables. **Extraneous variables** are other plausible explanations that could have brought about the observed changes in the outcome variable. For example, suppose students in a class using a new reading method score higher on a reading test than students using the current method. The researchers have to confirm that the higher scores are due to the method, rather than other extraneous, confounding variables, such as the teaching expertise of the experimental group teacher, amount of time devoted to reading, or ability levels of the experimental group of students. Prior to starting the study, the researchers have to review and control all possible extraneous variables that might affect the outcomes of the study. In our example, the researchers may want to ensure that both groups—experimental and control—are similar to each other before the new reading method is implemented. The researchers have to verify that both groups have capable teachers, have the same number of hours devoted daily or weekly to reading instruction, and that the reading ability of students in both groups is similar. When the extraneous variables are controlled, it is assumed that the groups differ from each other on one variable only—the reading instruction method. If experimental group students score higher on a reading test at the end of the study, the researchers can conclude that the new reading method is effective.

At times, extraneous variables develop during the study and are unforeseen. When researchers observe unexpected outcomes at the end of their study, they may want to probe further to determine whether some unplanned, extraneous variables are responsible for those outcomes. Often, when the research findings do not support the hypotheses stated at the beginning of their study, the researchers examine the design of the study to determine if any extraneous variables are responsible. When reporting their results, researchers are likely to include a discussion of possible extraneous variables in order to explain why their hypotheses were not confirmed.

A study is said to have a high **internal validity** when the researchers control the extraneous variables and the only obvious difference between the experimental and control groups is the intervention (that is, the independent variable). It makes sense, then, that a well-designed experiment has to have high internal validity to be of value. When there are uncontrolled, extraneous variables, they present competing explanations that can account for the observed changes in the dependent variable. One way to eliminate threats to internal validity and increase internal validity is to conduct studies in a lab-like setting under tight control of extraneous variables. Doing so, though, decreases the study's external validity. **External validity** refers to the extent to which the results of the study can be generalized and applied to other settings, populations, and groups. Clearly, if researchers want to contribute to their field (for example, education), their studies should have high external validity. The problem is that when studies are tightly controlled in order to have *high* internal validity, they tend to have *low* external validity. Thus, researchers have to strike a balance between the two. First and foremost, every study should have internal validity. But, when researchers control the study's variables too much, the study deviates from real life, decreasing the likelihood that the results can be generalized and applied to other situations. Since experimental studies must have internal validity to be of value, a brief discussion of the major threats to internal validity is presented next.[2]

Threats to Internal Validity

1. *History* refers to events that happen while the study takes place that may affect the dependent variable. For example, suppose a middle scool teacher wants to study the effect of a new instructional method (the independent variable) he is using this year to teach the Constitution to eighth-grade students. The dependent variable is the students' performance on the U.S. Constitution examination administered to eighth-grade students. The scores of students from the previous year are compared to those of this year's students who are using the new method. However, during this year the country is involved in a bitter, bipartisan political fight prior to the presidential elections. Assume further that this year's students score higher on the Constitution examination compared with last year's students. When the teacher evaluates the effectiveness of the new method, he should consider the effect of history as a possible threat to the internal validity of his study. He should confirm that the higher scores on the Constitution examination are due to the planned intervention (that is, the new instructional method) and not to the political events that have occurred while the study was going on.

---

2. HINT: For more information about threats to internal and external validity, you may want to review a book by Campbell and Stanley that, in spite of being old, is considered the most-cited source on the topic of experimental designs. Campbell, D. T., & Stanley, J. C. (1971). *Experimental and quasi-experimental designs for research.* Chicago: Rand McNally.

2. *Maturation* refers to physical, intellectual, or mental changes experienced by participants while the study takes place. Maturation is a particular threat to internal validity in studies that last for a longer period of time (as opposed to studies of short duration) or in studies that involve young children who experience rapid changes in their development within a short period of time. For example, suppose researchers want to enhance the fine motor coordination of preschoolers by providing time each day for them to practice tying their shoes. Before and after a five-month program, the children's coordination is tested. A significant improvement in the children's skills in tying their shoes may be due to the intervention (practice time). It is also possible that the children are better able to perform certain tasks that require fine motor coordination simply because they are older and their fine motor skills have developed over time.

3. *Testing* refers to the effect that a pretest has on the performance of people on the posttest. For example, in a study designed to test a new spelling method, students are asked to spell the same twenty-five words before and after the new instructional method is used. If they score higher on the posttest, it may be simply because they were exposed to the same words before rather than due to the effectiveness of the new method.

4. *Instrumentation* refers to the level of reliability and validity of the instrument being used to assess the effectiveness of the intervention. For example, in a study designed to assess the effectiveness of a new health education curriculum, a district-wide health test is used as the dependent variable. If the teacher finds that the scores of this year's students are not higher than last year's students, it may not be an indication that the curriculum is ineffective. Rather, it may be that the existing test does not measure the new skills and content emphasized by the new curriculum. In other words, the test lacks in validity and does not match the new curriculum.

5. *Statistical regression* refers to a phenomenon whereby people who obtain extreme scores on the pretest tend to score closer to the mean of their group upon subsequent testing, even when no intervention is involved. For example, suppose an IQ test is administered to a group of students. A few weeks later, the same students are tested again, using the same test. If we examine the scores of those who scored at the extreme (either very high or very low) when the test was administered the first time, we would probably discover that many low-scoring students score higher the second time around, while many high-scoring students score lower. The statistical regression phenomenon can pose a threat to internal validity in certain studies where the following occur: participants are selected for the study based on the fact that they have scored either very high or very low on the pretest, and the participants' scores on the posttest serve as an indicator of the effectiveness of the intervention.

6. *Differential selection* may be a threat in studies where volunteers are used in the experimental groups or in studies where preexisting groups are used as comparison groups (for example, one serves as experimental group and one as control group). This phenomenon refers to instances where the groups being compared differ from each other on some important characteristics even before the study begins. For example, if the experimental group is comprised of volunteers, they may perform better because they are high achievers and are highly motivated, rather than as a result of the planned intervention.

Threats to External Validity

Threats to external validity may limit the extent to which the results of the study can be generalized and applied to populations that are not participating in a study. It is easy to see why people might behave differently when they are being studied and observed. For example, being pretested or simply tested several times during the study may affect people's performance and motivation. Another potential problem could arise in studies where both experimental and control groups are comprised of volunteers and, therefore, may not be representative of the general population. Other potential problems that can pose a threat to external validity include: (a) people may react to the personality or behavior of those who observe or interview them; (b) people could try harder and perform better when they are being observed or when they view the new intervention as positive even before the start of the study; and (c) researchers may be inaccurate in their assessments when they have some prior knowledge of the study's participants before the start of the study.

Two well-known examples serve to illustrate some potential threats to external validity. One is called the **Hawthorne Effect**, named after a series of experimental studies conducted in 1924–1932 in a Western Electric Company plant in Hawthorne, Illinois, near Chicago. In the studies, researchers wanted to assess the effect of light intensity on workers' productivity. When the researchers increased the light intensity, it resulted in an increase in productivity. To confirm that the change in light intensity was indeed the cause for the increased productivity, the researchers *decreased* the light intensity. They discovered that productivity still went up. This experiment led them to conclude that the reason productivity went up in the first place was not the change in light intensity but the people's perception that they were being studied, which resulted in an increase in their level of motivation. Today, when designing experiments, researchers take into consideration the Hawthorne Effect, whereby the study's participants may behave in a certain way not necessarily because of the planned intervention but rather as a result of their knowledge that they are being observed and assessed.

Another threat is called the **John Henry Effect**. John Henry worked for a railroad company during the 1870s when the steam drill was introduced with the intention of replacing manual labor. John Henry, in his attempt to prove that men can do a better job than the steam drill, entered into a competition with the machine. He did win, but dropped dead at the finish line. Today, the John Henry Effect refers to conditions where control group members perceive themselves to be in competition with experimental group members and therefore perform above and beyond their usual level. In a study where the performance of control and experimental groups are compared, an accelerated level of performance of control group members may mask the true impact of the intervention on the experimental group members.

Comparing Groups

In conducting experimental research, the effectiveness of the intervention (the independent variable) is assessed via the dependent variable (the measured outcome variable). In all experimental studies, a posttest is used as a measure of the outcome, although not all studies include a pretest. Researchers try to compare groups that are as similar as possible prior to the start of the study so that any differences observed on the posttest can be attributed to the intervention. One of the best ways to create two groups that are as similar as possible is by *randomly* assigning people to the groups. Groups that are formed by using random assignment are considered similar to each other, especially when the group size is not too small.[3] When the groups being compared are small, even though they may have been created through random assignment, they are likely to differ from each other. Also, keep in mind that in real life, researchers are rarely able to randomly assign people to groups and they often have to use existing, intact groups in their studies.

Another approach that is used by researchers to create two groups that are as similar as possible is *matching*. In this approach, researchers first identify a variable that they believe may affect, or be related to, people's performance on the dependent variable (for example, the posttest). Pairs of people with similar scores on that variable are randomly assigned to the groups being compared. For example, two people who have the same verbal aptitude score on an IQ test may be assigned to the experimental or control group in a study that involves a task where verbal ability is important. The limitations of matching groups include the fact that the groups may still differ from each other because we cannot match them based on more than one or two variables. Also, there is a good possibility that we would end up with a smaller sample size because we would need to exclude from our study those people for whom we cannot find another person with a matching score.

Studies that involve a change in policy or a significant appropriation of money are likely to place a heavy emphasis on *scientifically based research* and on the use of experimental and quasi-experimental designs. When possible, the use of multiple sites, random assignments, or a careful matching of experimental and control groups is highly recommended to minimize threats to internal validity.

---

3. HINT: A rule of thumb followed by most researchers recommends a group size of at least thirty for studies where statistical tests are used to analyze numerical data.

Experimental group designs can be divided into three categories: *preexperimental,* *quasi-experimental,* and *true experimental* designs. The three categories differ from each other in their level of control and in the extent to which the extraneous variables pose a threat to the study's internal validity. In studies classified as preexperimental and quasi-experimental, when groups are compared, researchers may still not be able to confirm that differences between the groups on the posttest are caused by the intervention. The reason is that these studies involve the comparison of intact groups, which may not be similar to each other at the beginning of the study.

Studies using **preexperimental designs** do not have a tight control of extraneous variables, thus their internal validity cannot be assured. That is, researchers using these designs cannot safely conclude that the outcomes of the studies are due to the intervention. Studies using preexperimental designs either do not use control groups or, when such groups are used, no pretest is administered. Therefore, researchers cannot confirm that changes observed on the posttest are truly due to the intervention.

Studies using **quasi-experimental designs** have better control than those in preexperimental designs; nevertheless, there are still threats to internal validity, and potential extraneous variables are not well controlled in such studies. In quasi-experimental designs, the groups being compared are not assumed to be equivalent at the beginning of the study. Any differences observed at the end of the study may not have been caused by the intervention but are due to preexisting differences. It is probably a good idea to acknowledge these possible preexisting differences and to try to take them into consideration while designing and conducting the study, as well as when analyzing the data from the study. Studies using quasi-experimental designs include *time series* and *counterbalanced* designs. In **time-series designs**, groups are tested repeatedly before and after the intervention. In **counterbalanced designs**, several interventions are tested simultaneously, and the number of groups in the study equals the number of interventions. All the groups in the study receive all interventions, but in a different order.

True experimental designs offer the best control of extraneous variables. In true experimental designs, participants are randomly assigned to groups. Additionally, if at all possible, the study's participants are drawn at random from the larger population before being randomly assigned to their groups. Since the groups are considered similar to each other when the study begins, researchers can be fairly confident that any changes observed at the end of the study are due to the intervention.

Comparing Individuals

While most experimental studies involve groups, a growing number of studies in education and psychology focus on individuals. These studies, where individuals are used as their own control, are called **single-case** (or **single-subject**) **designs**. In these studies, individuals' behavior or performance is assessed during two or more phases, alternating between phases with or without an intervention. The measure used in single-case studies (that is, the dependent variable) is collected several times during each phase of the study to ensure its stability and consistency. Since the measure is used to represent the target behavior, the number of times the target behavior is recorded in each phase may differ from one study to another.

One of the most common single-case designs involves three phases and is called the **A-B-A single-case design**. The letter *A* is used to indicate the baseline phase where no intervention is applied, and the letter *B* is used to indicate the intervention phase.

The study begins by collecting several measures of the target behavior to establish the baseline (phase A). Then an intervention is introduced, during which the same target behavior is again measured several times (phase B). Next, the intervention is withdrawn and the target behavior is assessed again (phase A). The target behavior is compared across all phases, *with* the intervention (phase B) and *without* the intervention (phase A) to determine if the intervention was effective. If the target behavior is improved during phase B, the researchers can speculate that this was caused by the intervention. To rule out any extraneous variables as the possible cause for the change in the target behavior, it is assessed again during the withdrawal phase (the third phase). The expectation is that since no intervention is used at this phase (the second phase A), the target behavior should return to its original level at the start of the study (phase A). If the target behavior improves even though the intervention has been withdrawn, we may speculate that the intervention has produced a long-term positive effect. Studies may include repeated cycles of the baseline (phase A), treatment (phase B), withdrawal of treatment (return to baseline A), and treatment (phase B).

Basic single-case designs can be modified to include more than one individual and more than one intervention in the same study. Another modification to this design is in studies in which several individuals are studied simultaneously and the length of time of the baseline and treatment phases (phases A and B) differs from one person to another.

There are several potential problems associated with single-case studies. Because only one or a few individuals are studied, the external validity of the study (the extent to which the results can be generalized to other populations and settings) may be limited. To overcome this problem, single-case studies should be replicated. Another problem involves the nature of the intervention that may be used in single-case studies. In certain cases, it is not possible for researchers to withdraw the intervention and return to the baseline phase. For example, in a study where the intervention includes the implementation of new learning strategies, the teacher may not be able to tell the students during the withdrawal phase not to use these strategies once they have mastered them in the intervention phase. In other cases, withdrawing the intervention may pose ethical dilemmas. If the results from phase B convince the researchers that the treatment is effective, they may be reluctant to withdraw it in order to return to the baseline phase.

Note that single-case designs that use quantitative data are different from case studies that are used extensively in qualitative research. In the latter type, one or several individuals or "cases" (such as a student, a classroom, or a school) are studied in depth, usually over an extended period of time. Researchers employing a qualitative case study approach typically use a number of data collection methods (such as interviews and observations) and collect data from multiple data sources. They study people in their natural environment and try not to interfere or alter the daily routine. Data collected from these nonexperimental studies are usually in narrative form. In contrast, single-case studies, which use an experimental approach, collect mostly numerical data and focus on the effect of a single independent variable (the intervention) on the dependent variable (the outcome measure).

Nonexperimental Research

As mentioned before, nonexperimental research may be divided into two categories: causal comparative and descriptive. Causal comparative studies are designed to investigate cause-and-effect relationships without manipulating the independent variable. Descriptive studies simply describe phenomena.

Causal Comparative (Ex Post Facto) Research

In studies classified as causal comparative, researchers attempt to study cause-and-effect relationships. That is, they study the effect of the independent variable (the "cause") on the dependent variable (the "effect"). Unlike in experimental research, the independent variable is not being manipulated because it has already occurred when the study is undertaken, or it cannot or should not be manipulated. The following examples are presented to illustrate these points.

Let's say researchers want to study the effect of divorce on the parenting skills of individuals whose own parents had been divorced. The independent variable (that is, the parents' divorce) had occurred prior to the start of the study and therefore cannot be manipulated. Another example of causal comparative research may be a study designed to assess the effect of students' gender on their attitudes toward mathematics. Obviously, the independent variable (that is, the students' gender) is predetermined and cannot be manipulated.

At times, the independent variable can be manipulated, but researchers would not do so due to ethical reasons. For example, based on empirical data, researchers may speculate that children born to mothers who abuse drugs while pregnant are more likely to have learning disabilities in school compared with children of mothers who did not use drugs. However, the relationship between mothers' drug abuse and children's learning disabilities cannot be studied using experimental design. For obvious reasons, researchers are not going to randomly select a group of pregnant mothers and assign half of them to serve as the experimental group that is then told to use drugs.

Descriptive Research

Many studies are conducted to describe existing phenomena. Although researchers may construct new instruments and procedures to gather data, there are no planned interventions and no changes in the routine of people or phenomena being studied. The researchers simply collect data and interpret their findings. Thus, it is easy to see why qualitative research is considered nonexperimental.

Quite often, researchers conducting descriptive research use questionnaires, surveys, and interviews. The census survey, for example, is a nonexperimental study. Other examples of findings from a nonexperimental study include information presented in a report card—either that of an individual student or that of a school or a district. Information provided by governmental offices, such as the Consumer Price Index (CPI), is also based on nonexperimental research. Descriptive statistics may be used to analyze numerical data derived from nonexperimental studies. For example, a district may compare mean scores on standardized achievement tests or mobility rates from all schools in the district.

Correlation is often used in descriptive research. In most studies using correlation, a group of individuals is administered two or more measures and the scores of the individuals on these measures are compared. (For a discussion of correlation, see chapter 8.) For example, members of a school board may want to correlate scores on a norm-referenced achievement test of all fifth grade students in the district with their scores on a state-mandated achievement test that they have to administer to fifth graders each spring. If the correlation of the two tests is high, the school board members may propose that the district stop administering the norm-referenced test in fifth grade because similar information about the students' performance can be obtained from their scores on the state-mandated achievement test.

When researchers want to study how individuals change and develop over time, they can conduct studies using *cross-sectional* or *longitudinal* designs. In **cross-sectional designs**, similar samples of people from different age groups are studied at the same point in time. It is assumed that the older groups accurately represent the younger groups when they would reach their ages. Each person in the different samples is studied one time only. The biggest advantage of this design is that it saves time because data can be collected quickly. The biggest disadvantage is that we are studying different cohorts rather than following the same group of individuals. As an example, let's say that researchers want to study the physical and social development of preschool children from ages three to five years. Random samples of preschoolers ages three, four, and five are chosen and studied. Such a study is based on the assumption that the children who are three at the time the study is conducted would behave the following year like the children in the study who are currently four, and that those who are currently four would behave the following year like those who are five at the time the study is conducted.

Longitudinal studies are used to measure change over time by collecting data at two or more points for the same or similar groups of individuals. The greatest advantage of this design is that the same or similar individuals are being followed; a major disadvantage is that the study lasts a long time. There are three types of longitudinal studies: *panel, cohort,* and *trend.*

In a **panel study**, the same people are studied at two or more points in time. For example, in 1962, the first group of preschoolers entered the Head Start program in Ypsilanti, Michigan. Children in the program were followed for many years. Another example is a study that began in 1921 by Lewis Terman and his associates at Stanford. Using surveys and interviews, the researchers followed a group of about 1,500 gifted boys and girls between the ages of three and nineteen, with IQ scores above 135, as these boys and girls aged and matured.

In a **cohort study**, similar people, selected from the same cohort, are studied at two or more points in time. For example, a university may survey its students to compare their attitudes toward the choice of classes offered to them. In the first year, a group of freshmen is selected and surveyed. In the second year, a group of sophomores is selected and surveyed. The following year, a group of juniors is selected and surveyed, followed by a group of seniors the next year.

In a **trend study**, the same research questions are posed at two or more points in time to similar individuals. For example, teachers and other educators may be asked for their opinions about homeschooling every year or every five years to allow researchers to record and note any trends and changes over time.

SUMMARY

1. **Research** is a systematic inquiry that includes data collection and analysis. The goal of research is to describe, explain, or predict present or future phenomena. In the context of education, these phenomena are most likely to be behaviors associated with the teaching and learning processes.
2. **Basic research**, whose main goal is to develop theories and generalizations, is conducted mostly in labs, under tightly controlled conditions.
3. **Applied research** is aimed at testing theories and applying them to specific situations.
4. **Action research** is conducted by practitioner-researchers in their own settings to solve a problem by studying it, proposing solutions, implementing the solutions, and assessing the effectiveness of these solutions.

5. The terms **practitioner research**, *classroom research*, and *teacher-as-researcher* are often used in place of the term *action research*.
6. **Quantitative research** is often conducted to study cause-and-effect relationships and to focus on studying a small number of variables and collecting numerical data.
7. **Qualitative research** seeks to understand social or educational phenomena and focuses on one or a few cases that are studied in depth using multiple data sources. Data sources used in qualitative research are subjective in nature (for example, interviews and observations), and they yield mostly narrative data.
8. In many textbooks, quantitative research is equated with experimental designs and the use of numerical data, while qualitative research is equated with descriptive research and narrative data.
9. A better way to describe quantitative and qualitative paradigms is to state that the paradigms are two endpoints on a continuum. Studies can be located at different points along this continuum.
10. While most experimental studies use numerical data and most descriptive studies use narrative data, both numerical and narrative data can be used in experimental or descriptive studies.
11. When educational practitioners conduct research, they often use a small sample size and employ a case-study approach.
12. In experimental research, researchers plan an *intervention* and study its effect on groups or individuals. The intervention is also called the **independent variable** or *treatment*.
13. Experimental research is designed to test the effect of the independent variable on the outcome measure, called the **dependent variable**.
14. **Nonexperimental research** may be divided into *causal comparative* (also called *ex post facto*) research and *descriptive research*.
15. **Descriptive research** is aimed at studying phenomena as they occur naturally without any intervention or manipulation of variables.
16. In many experimental studies, the experimental group that receives the treatment (the intervention) is compared to the control group that receives no treatment or is using the existing method. In other experimental studies, the performance of the same group is compared before the intervention (the pretest) and after the intervention (the posttest).
17. **Extraneous variables** are variables—other than the planned intervention—that could have brought about changes that are measured by the dependent variable. Extraneous variables may be unforeseen and develop during the study, especially when the study lasts for a long period of time.
18. A study is said to have high **internal validity** when the extraneous variables are controlled by the researchers and the only obvious difference between the experimental and control groups is the planned intervention (that is, the independent variable).
19. **External validity** refers to the extent to which the results of the study can be generalized and applied to other settings, populations, or groups.

20. Threats to internal validity include: **history, maturation, testing, instrumentation, statistical regression**, and **differential selection**.
21. Threats to external validity include the **Hawthorne Effect** and the **John Henry Effect**.
22. Experimental group designs can be described as *preexperimental, quasi-experimental*, and *true experimental*.
23. **Preexperimental designs** do not have tight control of extraneous variables, and their internal validity is low.
24. In **quasi-experimental designs**, the groups being compared are not assumed to be equivalent prior to the start of the study. Studies in this category have better control of extraneous variables compared with preexperimental designs. Examples of quasi-experimental designs are *time-series* and *counterbalanced designs*.
25. In **time-series designs**, groups are tested repeatedly before and after the intervention.
26. In **counterbalanced designs**, several interventions are tested simultaneously and the number of groups in the study equals the number of interventions. All the groups in the study receive all interventions, but in a different order.
27. The most important aspect of **true experimental designs** is that participants are assigned at random to groups. Therefore, the groups are considered similar to each other at the start of the study.
28. In **single-case** (or **single-subject**) **designs**, the behavior or performance of people is assessed during two or more phases, alternating between phases with and without intervention.
29. The measure (that is, the dependent variable) used in a single-case study is administered several times during each phase of the study to ensure the stability and consistency of the data.
30. One of the most common single-case designs is the **A-B-A single-case design**. In studies using this design, the target behavior (that is, the dependent variable) is measured before the intervention (phase A), during the intervention (phase B), and during the second phase A, when the intervention is withdrawn.
31. A single-case design can be modified to include more than one individual and more than one intervention in the same study.
32. Qualitative research is considered nonexperimental, and many researchers conducting nonexperimental research use qualitative approaches and collect narrative data.
33. Researchers conducting descriptive research may collect narrative or numerical data. The census is an example of a descriptive study where a large amount of numerical data is collected.
34. Correlation is a statistical test often used in descriptive research. In most correlational studies, two or more measures are administered and participants' scores on these measures are compared. (See chapter 8.)
35. When researchers want to study how individuals change and develop over time, they may use *cross-sectional* or *longitudinal designs*.

36. In **cross-sectional studies**, similar samples of people from different age groups are studied at the same point in time.
37. **Longitudinal studies** are used to measure change over time by collecting data at two or more points for the same or similar groups of individuals over a period of time. There are three types of longitudinal studies: *panel*, *cohort*, and *trend*.
38. In a **panel study**, the same people are studied at two or more points in time. In a **cohort study**, similar people, selected from the same cohort, are studied at two or more points in time. In a **trend study**, the same research questions are posed to similar individuals at two or more points in time.

CHECK YOUR UNDERSTANDING

1. Design an *experimental* study in your educational setting. This should include a *group* of participants.
    a. Why do you consider it an *experimental* study? Explain.
    b. What would be the *independent* and *dependent* variable(s)?
    c. What might be some threats to the study's *internal* validity?
2. Design an *experimental* study of *one* or a *few individuals* in your setting.
    a. What would be the purpose of the study?
    b. What data would you collect? How would you use the data?
    c. To whom will you be able to generalize the findings?
3. Design a *nonexperimental* study in your educational setting of a *group* of participants.
    a. What would be the purpose of the study?
    b. What type(s) of data would you collect and why?

Basic Concepts in Statistics

Chapter 2 introduces to you several basic concepts in statistics that are referred to and used in other chapters in the book. Each concept is defined and explained, and concrete examples are provided to further illustrate the concepts. The major topics covered in this chapter include: *variables* and *measurement scales*, *population* and *sampling*, and *parameters* and *statistics*. The use of *hypotheses* in the process of statistical testing is highlighted. You will also learn how to evaluate your statistical results and how to decide if they confirm your hypotheses.

You will probably find yourself referring back to certain sections of this chapter as you learn about the various statistical tests that are presented in this book. For example, the process of stating hypotheses or deciding whether they are confirmed is an integral part of several statistical tests discussed in the book (for example, correlation and *t* test). Therefore, this chapter is an important introduction to other chapters in the book.

The term **statistics** refers to methods and techniques used for describing, organizing, analyzing, and interpreting numerical data. Statistics are used by researchers and practitioners who conduct research in order to describe phenomena, find solutions to problems, and answer research questions.

VARIABLES AND MEASUREMENT SCALES

A **variable** is a measured characteristic that can assume different values or levels. Some examples of variables are age, grade level, height, gender, and political affiliation. By contrast, a measure that has only one value is called a **constant**. For example, the number of hours in a day—24—is also a constant. Or, the length of each side of a square with a perimeter of 12 inches is a constant; that is, all sides are equal (as opposed to other geometric shapes where the sides may have different lengths).

The decision about whether a certain measure is a constant or a variable may depend on the purpose and design of the study. For example, grade level may be a variable in a study where several grade levels are included in an attempt to measure a particular type of growth over time (for example, cognitive abilities or social skills). In a study where three different instructional methods are used with first graders to see which method is preferred by the students, grade level is a constant.

A variable may be continuous or discrete. **Continuous variables** can take on a wide range of values and contain an infinite number of small increments. Height, for example, is a continuous variable. Although we may use increments of one inch, people's heights can differ by a fraction of an inch. **Discrete variables** contain a finite number of distinct values between any two given points. For example, on a classroom test, a student may get a score of 80 or 81, but not a score of 80.5.

It is important to remember that in the case of intelligence tests, for instance, while the test can only record specific scores (discrete variable), intelligence itself is a continuous variable. Then again, research reports may describe discrete variables in the manner usually prescribed for continuous variables. For example, in reporting the number of children per classroom in a given school, a study may indicate that there are, on average, 26.4 children per room. In actuality, the number of children in any given classroom would, of necessity, be indicated by a whole number (for example, 26 or 27). Reporting the discrete variable in this manner lets the researcher make finer distinctions.

Measurement is defined as assigning numbers to observations according to certain rules. Measurement may refer, for example, to counting the number of times a certain phenomenon occurs or the number of people who responded "yes" to a question on a survey. Other examples include using tests to assess intelligence or to measure height, weight, and distance. Each system of measurement uses its own units to quantify what is being measured (for example, meters, miles, dollars, percentiles, frequencies, or text readability level).

There are four commonly used types of measurement scales: *nominal, ordinal, interval,* and *ratio.* For all four scales we use numbers, but the numbers in each scale have different properties and should be manipulated differently. It is the duty of the researcher to determine the scale of the numbers used to quantify the observations in order to select the appropriate statistical test to analyze the data. Further explanation of the four scales is presented next.

Nominal Scale

In **nominal scales**, numbers are used to label, classify, or categorize data. For example, the numbers assigned to the members of a football team comprise a nominal scale, where each number represents a player. Numbers may also be used to describe a group in which all members have some characteristic in common. For example, in coding data from a survey to facilitate computer analysis, boys may be coded as "1" and girls as "2." In this instance, it clearly does not make sense to add or divide the numbers. We cannot say that two boys, each coded as 1, equal one girl, coded as 2, although in other contexts, $1 + 1 = 2$. Similarly, it will not make sense to report that the average gender value is, for example, 1.5! For nominal scales, the numbers are assigned arbitrarily and are interchangeable. Consequently, instead of assigning 1 to boys and 2 to girls, we can just as easily reverse this assignment and code boys as 2 and girls as 1.

Ordinal Scale

For **ordinal scales**, observations can be ordered based on their magnitude or size. This scale has the concept of *less than* or *more than.* For example, using grade point average (GPA) as a criterion, a student who is ranked tenth in the class has a higher GPA than a student who is ranked fifteenth. But we do not know how many GPA points separate these two students. The same can be said about three medal winners in the long jump at the Olympic Games. It is clear that the gold medalist performed better than the silver medalist, who, in turn, did better than the bronze medalist. But we should not assume that the same number of inches separate the gold medalist from the silver medalist as those inches separating the silver medalist from the bronze medalist. Thus, in an ordinal scale, observations can be rank-ordered based on some criterion, but the intervals between the various observations are not assumed to be equal.

Interval Scale

Interval scales have the same properties as ordinal scales, but they also have equal intervals between the points of the scale. Temperatures, calendar years, IQ scores, and achievement test scores all are considered interval scales. The difference between a temperature of 20°F and 25°F is 5°F, and is the same as, let's say, the difference between 65°F and 70°F. However, we cannot say that a temperature of 90°F is three times as hot as a temperature of 30°F or that a child with an IQ of 150 is twice as smart as a child with an IQ of 75 because an interval scale does not have an absolute, or true, zero. An absolute zero is a point lacking completely the characteristic being measured. In Fahrenheit temperature, the temperature of 0° does not imply lack of heat. (The absolute zero is –273°F, where the molecules do not move at all.) Similarly, the zero point in an IQ scale is not a true zero, because we cannot say that a person who received a score of zero on our IQ test has no intelligence at all. We probably can find other questions that this person can answer, but these questions were not asked. Most of the numerical examples used in this book are measured using an interval scale.

Ratio Scale

Ratio scales have the same characteristics as interval scales, but in addition they have an absolute zero. Thus, we can compare two points on the scale and make statements such as "This point is twice as high as that point," or "This person is working half-time" (as opposed to full-time). Height, for example, is a ratio scale. We can say that a person whose height is 3'2" is half as tall as a person whose height is 6'4". Height has a true zero point, usually the floor on which the person stands while being measured. Or, for example, in a race, the absolute zero point is when the gun sounds and the stopwatch is pressed to start the counting. Ratio scales exist most often in the physical sciences but rarely in behavioral sciences (such as education, psychology, or sociology).

POPULATIONS AND SAMPLES

An entire group of persons or elements that have at least one characteristic in common is called a **population**. Examples would include all the residents of a particular suburb, all high school mathematics teachers in a district, or all the states in the United States. A population may have more than one characteristic, or trait, in common. For example, we may talk about a population of female students in the local state university who are majoring in computer sciences.

In real life, we rarely study and measure entire populations. The most notable example of a study of the entire population is the census, which is conducted once every ten years. Clearly, including *all* members of a population in a study is expensive, time-consuming, and often simply impractical. Yet, most research studies are concerned with generalization and obtaining findings that describe large groups. Thus, quite often, researchers draw a sample and use it to gain information about the population. A **sample**, then, is a small group of observations selected from the total population. A sample should be *representative* of the population because information gained from the sample is used to estimate and predict the population characteristics that are of interest.

As an example, suppose we want to know what the parents of the students in the elementary school district think about sex education. In a large district, we may have as many as 10,000 parents, and it might be too expensive to survey every household. Instead, a sample of 500 parents who represent the various grade levels and schools in the district may be surveyed. The results of this survey can be said to represent all the parents in the district. Of course, as with every survey, the response rate has to be adequate to ensure that the results truly reflect the total population.

A certain level of error is expected when we use samples to estimate populations. Some chance variation in the sample numerical values (for example, mean) occur when we repeatedly select same-size samples from the same population and compare their numerical values. This error, called a **sampling error**, is beyond the control of the researcher.

PARAMETERS AND STATISTICS

A measure that describes a characteristic of an entire population is called a **parameter**. The number of school-age children in the district who get free or reduced lunch is a parameter because it describes a certain characteristic of the entire district. A **statistic** is a measure that describes a characteristic of a sample.[1]

In most research studies, we are interested in obtaining information about a population parameter. But, instead of obtaining the parameter directly by measuring every member of the population, we draw a sample, measure the sample to obtain the statistic, and then use that statistic to estimate the corresponding population parameter.

For example, an electronic gaming company that develops computer action games may want to pilot test a new game designed for teenage children. The company may select a sample of teenagers, let them play with the new computer game, observe their reactions to the game, and ask them for their opinions. The company will then generalize the findings from the sample to the total population of teenagers who are the potential users of this new computer game.

Clearly, the sample selected should be representative of the population because the sample statistics are used to estimate and predict the population parameters. There are a number of procedures that may be used to select a sample. The next section describes several of the most common sampling techniques used by researchers.

METHODS OF SAMPLING

The majority of research studies in education are designed to study populations by using representative samples. In many studies in the physical sciences it is quite simple to obtain a representative sample. For example, assume that a scientist wants to study

---

1. HINT: Note the difference between two similar terms: *a statistic* and *statistics*. A statistic is a numerical index or value of a sample, whereas statistics refers to a group of methods, techniques, and analyses of numerical data.

the quality of the water of a pool. All the scientist needs to do is to scoop out a small amount of pool water and analyze this sample. We would all agree that the sample of water in the jar is representative of the pool water. In education, though, as well as in other behavioral sciences (for example, psychology and sociology), the task of obtaining a representative sample of babies, children, or adults is much more complicated.

There are several sampling methods that may be used in research. In choosing the sampling method, the researcher has to decide which one is appropriate and feasible in a given situation. Most sampling methods share the same steps, or sequence: first, the population is identified; then the sample size required is determined; and lastly, the sample is selected.

Simple Random Sample

In selecting a **simple random sample**, every member of the population has an equal and independent chance of being selected for inclusion in the sample. That is, the selection of one member in no way increases or decreases the chances of another member also being selected. Sampling procedures whereby the first 100 people who stand in line are chosen, or every other person from a list is selected, do not fit the definition of a random sample. When the first 100 people are selected, those who stand behind them do not have a chance of being included in the sample. Likewise, choosing every other person means that persons next to those being selected do not have a chance of being included.

In theory, if the random sample is large enough, it will truly represent the population in every respect and be an accurate reflection of the population. Selecting ten people from a population of 1,000, even if done by using a random sampling procedure, may result in a sample that is not truly representative of the population.

The idea that comes to mind when we think of a random sample is drawing names from a hat. While a random sample can be drawn this way, such a process is not efficient and there are more practical means of achieving the same results. For example, using the computer to draw a sample using a table of random numbers offers a faster way of drawing a random sample.

Systematic Sample

In a **systematic sample**, every *Kth* member (for example, every fifth or tenth person) is selected from a list of all population members. The procedure starts by ascertaining the size of the population and the desired sample size. The population size is then divided by the sample size to obtain the value of *K*. For example, if we have a population of 500 and need a sample of 25, we divide 500 by 25 to obtain a *K* of 20. In other words, we select every twentieth member to achieve the desired sample size of 25.

A systematic sample can be a good representation of the population when the names on the list from which the sample members are selected are listed randomly. Since this is rarely the case, the sample may be biased. For example, certain nationalities tend to have many last names that start with the same letter. Thus, a whole group of people of the same nationality may be excluded from the sample if the names of those selected are listed just before and just after that group of names.

Stratified Sample

To obtain a **stratified sample**, the population is first divided into subgroups (that is, *strata*), and then a random sample is taken from each subgroup. Stratified sampling is used extensively in market research, in political polls, and in norming standardized tests. The final sample represents, proportionately, the various subgroups in the population. A stratified sample may be used when there is a reason to believe that various subgroups in the population may have different opinions or behave differently because of some characteristics that the group members have in common. An example may help to illustrate this sampling procedure.

Suppose a large, urban elementary school district with 5,000 teachers wants to survey its teachers about their attitudes toward merit pay. Instead of surveying all 5,000 teachers, a stratified sample of 250 may be selected. The teachers may first be divided into strata based on variables such as grade level taught (primary, intermediate, or upper), subjects taught, tenure status, and annual salary (in increments of $5,000). For example, we may have a stratum of intermediate grades, social studies tenured teachers, whose annual salary is $80,000 to $85,000. From each subgroup, a random sample may be drawn. The resulting sample of 250 teachers will include, proportionately, all subgroups from the total district population of 5,000 teachers. Thus, the sample of teachers that will participate in the survey will be a miniature version of the population where each stratum is represented in proportion to its size in the population.

Convenience Sample

Occasionally, researchers conduct studies using an accessible sample, such as the researchers' own classrooms or schools. A **convenience** (or **incidental**) sample is a group that is chosen by the researcher to participate in the study because of its convenience. For example, college psychology professors may choose, for convenience and cost-saving purposes, to use their own students to conduct an experimental study. Or graduate students working on their theses or dissertations may use their own schools to collect data. In fact, a fair number of research studies in education and psychology are done using a convenience sample.

The main problem in using incidental sampling is that it is not always clear what population the sample belongs to, since the study did not start by identifying a population of interest and choosing a sample from that defined population. Great care should be exercised in generalizing the results of the study to a larger population.

SAMPLE BIAS

Sample bias refers to *systematic* (as opposed to *random*) differences between the sample and the population from which it was selected. If, for example, a political poll is conducted by randomly selecting and calling respondents using a database of cell phone numbers, the resulting sample is likely to be biased because it excludes voters who do not own cell phones or those whose numbers are unlisted.

A well-publicized example of such sample bias occurred during the 1936 presidential election when Republican Alf Landon ran against Democrat Franklin D. Roosevelt. The *Literary Digest* predicted a victory by Landon after receiving a 25 percent response rate from 10 million written ballots, which were mailed out. The mailing list for the ballots was based on telephone books and state registries of motor vehicles. Of course, such a list left out a big segment of the population that voted in the presidential election but was not included in the survey because they did not own a telephone or a car.

Another possible sample bias stems from using volunteers in a study. Even though the volunteers may come from a clearly defined population, they may not be "typical" of other members of that population. Conducting a study with volunteers and then generalizing the results to the population at large can lead to inaccurate conclusions.

A sample may also be biased when it is based solely on the responses of people who complete and send back the surveys. Those responding are often similar to people who volunteer to participate in a study; therefore, their responses may, or may not, represent the rest of the population. In many cases, those who respond to surveys feel strongly one way or another about the topic of the survey, whereas the majority of people do not bother to respond. Yet, quite often, the responses of those who return their surveys are generalized to the total population.

SIZE OF SAMPLE

As sample size increases, it is more likely to be representative of the population, especially when the sample is randomly selected. In well-designed experimental studies, a sample that is a smaller portion of the population may be sufficient to provide an accurate representation when chosen from larger population sizes. For example, when the population is greater than 10,000, a sample of 1,000 to 1,500 may give adequate precision.

In determining whether a sample truly represents the population, it is important to consider how the sample was selected as well as the size of the sample used. For example, a sample that is drawn by using a simple random sampling approach is highly regarded. Nevertheless, if that sample consists of only five students, it is probably not an adequate representation of the population from which it was selected. By the same token, size alone does not guarantee an accurate sample; a large sample may also be biased. In general, it is recommended that researchers try to obtain as large a sample as is feasible. A sample size of at least thirty cases or subjects is recommended in most studies in education.

PARAMETRIC AND NONPARAMETRIC STATISTICS

There are different research situations that call for the use of two types of statistics: *parametric* and *nonparametric*.[2] **Parametric statistics** are applied to data from populations that meet the following assumptions: (a) the variables being studied are measured on an interval or ratio scale, (b) individuals are randomly assigned to groups, (c) the scores are normally distributed, and (d) the variances of the groups being compared are similar. When these assumptions are met, researchers are likely to use parametric tests that are more efficient and powerful than their nonparametric counterparts. However, in many research situations in behavioral science and education, it is hard to meet all the required assumptions. As a result, findings should be interpreted cautiously. It is probably safe to say that many researchers always use interval or ratio scales when applying parametric tests, while it is more common for researchers to violate the other assumptions.

Nonparametric statistics are used with ordinal and nominal data, or with interval and ratio scale data that fail to meet the assumptions needed for parametric statistics. Nonparametric statistics are easier to compute and understand, compared with parametric statistics. The chi square test, for example (see chapter 11), is a nonparametric statistic, whereas the *t* test (see chapter 9) and analysis of variance (see chapter 10) are examples of parametric statistics. The majority of the statistical tests you are likely to read about in the literature are classified as parametric.

DESCRIPTIVE AND INFERENTIAL STATISTICS

The field of statistics is often divided into two broad categories: *descriptive statistics* and *inferential statistics*. **Descriptive statistics** classify, organize, and summarize numerical data about a particular group of observations. There is no attempt to generalize these statistics, which describe only one group, to other samples or populations. Some examples of descriptive statistics are the mean grade point average (GPA) of the students in the biology honors class, the number of students in the district, and the ethnic makeup of students at a given university.

Inferential statistics (which may also be called *sampling statistics*), involve selecting a sample from a defined population and studying that sample in order to draw

---

2. HINT: Nonparametric statistics are also called *assumption-free* or *distribution-free* statistics.

conclusions and make inferences about the population. The sample statistics are then used to estimate the population parameters. The rationale behind inferential statistics is that since the sample represents the population, what holds true for the sample probably also holds true for the population from which the sample was drawn.

In political polls, for example, a pollster may survey 1,500 voters and use their responses to predict the national election results the next day. Another example may be of a curriculum coordinator in a large high school district who is conducting a study to explore the efficacy of using cooperative learning in science classes and the effects of this approach on students' achievement and attitudes. Four teachers who are trained to use cooperative learning agree to pilot test this approach in their classes for one year. At the end of the year, the curriculum coordinator gathers students' achievement and attitudes scores. These scores are then used to decide whether to implement cooperative learning in all the high school science classes in the district.

Descriptive and inferential statistics are not mutually exclusive. In a sense, inferential statistics include descriptive statistics. When a sample is observed and measured, we obtain descriptive statistics for that sample. Inferential statistics can take the process one step further and use the information obtained from the sample to estimate and describe the population to which the sample belongs. Whether or not a given statistic is descriptive or inferential does not necessarily depend on the type of statistic, but rather on its purpose. On one hand, the mean score on a spelling test is a descriptive statistic if the teacher wants to compare the scores of the students on the test that was given on Friday to the scores of the same students on the same test given at the beginning of the week. On the other hand, the same spelling test can be given in another study done in the district. In that study, spelling scores from two randomly selected groups of second grade students are compared: the students in one group are using a new method to learn spelling, and the students in the other group are using the traditional spelling method. The results from the two instructional methods can be evaluated to determine if, indeed, the new method is more effective. If this were the case, the language arts coordinator in the district may recommend that other second grade teachers in the district use the new method.

USING HYPOTHESES IN RESEARCH

A research study often begins with a **hypothesis** (an "educated guess") that is a prediction about the outcome of the study. After the hypothesis is proposed, a study is designed to test that hypothesis. The data collected in the study enable the researchers to decide whether the hypothesis is supported. Hypotheses should be clearly and concisely stated and be testable.

A study may have more than one hypothesis. For example, in a study of middle school students, their attitudes toward school are assessed using a questionnaire and their school performance is measured using their GPA. One hypothesis in this study may predict that the girls' mean score on the questionnaire would be significantly higher than the boys' mean score, while another hypothesis may predict a positive correlation between students' scores on the questionnaire and their GPA.

Alternative and Null Hypotheses

Two types of hypotheses are used in research to explain phenomena and to make predictions about relationships between variables in a study. These two hypotheses are the *alternative hypothesis* and the *null hypothesis*. The **alternative hypothesis** (represented by H_A or H_1) guides the investigation and gives direction to the design of the study. Often, the alternative hypothesis is simply referred to as *the hypothesis* or *the research hypothesis*.[3] It usually predicts that there will be some relationship between variables, or difference between means or groups. For example, the alternative hypothesis may state that there will be a positive correlation between students' reading fluency and their reading comprehension scores. Or, the alternative hypothesis may predict that students in classes where the teachers use differentiated instruction will score significantly higher on the end-of-year spelling test compared with students in similar classes where teachers do not use differentiated instruction.

The **null hypothesis** (represented by H_O) usually predicts that there will be no relationship between variables or no difference between groups beyond that which may be attributed to chance alone. In most cases, the null hypothesis (which may also be called the *statistical hypothesis*) is not formally stated, but is always implied. The following two examples may illustrate how the null hypothesis is used in educational research. In the first example, we will conduct an experimental study to test the null hypothesis. This study would be conducted to test the effect of starting the school day half an hour later on students' achievement test scores. In one junior high school in the district, the students would start their day half an hour later and in the other school in the district, the students would continue with the same schedule as in past years. The null hypothesis in this study states that there would be no difference in the mean scores on an achievement test between the students in the two junior high schools who start school at different times. In our second example, the null hypothesis states that there would be no significant correlation between IQ and depression scores in college students. This hypothesis would be tested using a random sample of 200 students from one university. IQ and depression scores of those students would be obtained and correlated to test the null hypothesis.

Directional and Nondirectional Hypotheses

Hypotheses may be stated as *directional* or *nondirectional*. A **directional hypothesis** predicts the direction of the outcomes of the study. In studies where group differences are investigated, a directional hypothesis may predict which group's mean

---

3. HINT: In this book, whenever we use the word *hypothesis*, we are referring to the *alternative hypothesis*, whereas the *null hypothesis* is always called *null hypothesis*.

would be higher. In most experimental studies, researchers are likely to use a directional hypothesis that predicts differences in performance or behavior of experimental and control groups on the dependent variable. In other words, they are quite certain that there will be a difference between the groups as a result of the intervention. In studies that investigate relationships between variables, directional hypotheses may predict whether the correlation will be positive or negative.

A **nondirectional hypothesis** predicts that there will be a difference or relationship, but the direction of the difference or association is not specified. For example, the researcher predicts that one group's mean will be higher, but the hypothesis does not specify which one. Similarly, when the researcher predicts a statistically significant relationship but cannot predict whether the relationship would be positive or negative, the hypothesis is nondirectional.

PROBABILITY AND LEVEL OF SIGNIFICANCE

Statistical results from research studies may be used to decide whether to retain (that is, accept) or reject the null hypothesis. Once this first decision is made, a researcher can then determine whether the alternative hypothesis has been confirmed. It should be mentioned, though, that this statistical decision is made in terms of *probability*, not *certainty*. We cannot *prove* anything; we can only describe the probability of obtaining these results due to sampling error or chance. For example, we may want to compare the means from experimental and control groups using a statistical procedure called the *t* test for independent samples (see chapter 9). The null hypothesis states that the difference between the two means is zero. The statistical results may lead us to two possible conclusions:

1. It is unlikely that the two means came from the same population, and the difference between them is too great to have happened by chance alone. The null hypothesis is *rejected*.
2. The difference between the two means is not really greater than zero, and the two means probably came from the same population. In such cases, even if we observe some differences between the two means, we attribute them to sampling error and not to some systematic differences resulting from the experimental treatment. The null hypothesis is retained.

In most statistical tests, the probability level of 5 percent (*p* value of .05) serves as the cutoff point between results considered statistically significant and those considered not statistically significant.[4] The *p* **level** (that is, **level of significance**) indicates the *probability* that we are rejecting a true null hypothesis. Findings are usually reported as **statistically significant** if the probability level is 5 percent or less ($p \leq$

---

4. HINT: In statistics, the term *significant* does not necessarily mean the same as "useful in practice" or "important."

.05).[5] If the probability level is higher than 5 percent ($p > .05$) researchers are likely to report the findings as *not statistically significant*, rather than report the *actual p level*. As you read published research reports, you may find that researchers often list the exact probability level (p value) when their results are statistically significant instead of using the 5 percent cutoff point.

Regardless of the research hypothesis presented at the outset, the statistical testing and the evaluation of the findings start with a decision regarding the null hypothesis. To make a decision about the null hypothesis, we first calculate the sample statistic to get the obtained value. We then compare the *obtained* value to the appropriate *critical* value. The **critical value** determines the cutoff point between statistical results that are considered *statistically significant* and those that are *not statistically significant*. While these values can be determined from statistical tables of critical values (which are found often in the appendix of statistics textbooks), statistical software packages (such as IBM SPSS) provide the exact p value as part of their output. Therefore, in this book, we provide only short relevant sections from these tables, embedded into the text.

If the obtained value *exceeds* the critical value, the null hypothesis is *rejected*. *Rejecting* the null hypothesis means that the probability of obtaining these results by chance alone is very small (for example, less than 5 percent). We conclude that the relationship or difference, as predicted by the alternative hypothesis (H_A), is probably true. *Retaining* the null hypothesis means that these results (for example, the difference between two means) may be due to sampling error and could have happened by chance alone more than 5 percent of the time.

There is a clear relationship between the sample size and the confidence level in rejecting the study's null hypothesis. As the sample size *increases*, a *lower* computed test statistic value is needed in order to reject the null hypothesis at the $p = .05$ level. To illustrate this point, let's look at studies that use the Pearson correlation. The null hypothesis in such studies is that the correlation coefficient r is equal to zero ($r = 0$). (See chapter 7 for a discussion of correlation.) For example, with a sample size of 30 ($n = 30$), the correlation coefficient has to be at least .349 ($r = .349$) to be considered statistically significant at $p = .05$. As the sample size increases to 50 ($n = 50$), a correlation coefficient of $r = .273$ would be considered statistically significant at $p = .05$. And when the sample size is 100 ($n = 100$), a correlation coefficient as low as $r = .195$ is statistically significant at $p = .05$. Note the inverse relationship between the sizes of the samples and the magnitude of the correlation coefficients. Thus, with a very large sample size, even low correlation coefficients are going to be defined as statistically significant.

This book provides step-by-step explanations of the processes for determining the p values in each of the statistical test examples in the book (see chapters 7–11). In real

---

5. HINT: When the results are statistically significant, report the highest (the best) level of significance. For example, if results are significant at the $p < .01$ level, report that level, rather than $p < .05$. Of course, you can always report the exact p value (e.g., $p = .03$).

life, though, you will probably use a computer program to compute the p values. There are several powerful computer software programs (such as SPSS) readily available to novices, as well as experienced researchers.[6] These programs can analyze your statistical data and will, in most cases, provide the exact p values.

ERRORS IN DECISION MAKING

When the probability level is set at the beginning of the study, *before* collecting and analyzing data, it is represented by the Greek letter alpha (α). The convention is to use an alpha of .05. Nevertheless, in some exploratory studies, researchers may set alpha at .10. In other studies, the researchers may want to set the alpha level at .01 so as to have a higher level of confidence in their decision to reject the null hypothesis.

When researchers decide to *reject* the null hypothesis (H_O) when in fact it is true and should not be rejected, they are making a **Type I error**. And, when they decide to *retain* the null hypothesis when in fact it should be rejected, they are making a **Type II error**. The proper decision is made when researchers *reject* a *false* null hypothesis or when they *retain* a *true* null hypothesis.

If we decide to set alpha at $p = .01$ (instead of .05), we *decrease* the chance of making a Type I error because we are less likely to reject the null hypothesis. Conversely, in setting alpha at .01, we *increase* the chance of making a Type II error and are more likely to retain the null hypothesis when in fact we should have rejected it.

DEGREES OF FREEDOM

In order to consult tables of critical values, the researcher needs to know the **degrees of freedom** (**df**). Essentially, *df* is calculated as $n - 1$ (the number of cases or subjects in the study minus 1), although there are some modifications to this rule in some statistical tests. The exact way to calculate the *df* will be explained in the discussion of each of the statistical tests that are included in this book. Note that computer software programs will compute the degrees of freedom for you so you do not have to worry about the exact way to calculate the *df* in your study.

EFFECT SIZE

The decision whether to retain or reject the null hypothesis is affected greatly by the study's sample size. A large sample size may lead researchers to reject the null hypothesis even when there are very small differences between the variables or when the correlation between the variables is very low. Conversely, in studies where a small sample size is used, researchers may decide to retain the null hypothesis even when there are large differences or a high correlation between variables. In the last twenty years, the concept of *effect size* has gained much popularity as another way to evaluate the statistical data gathered in research studies. The American Psychological Associa-

---

6. HINT: Excel can also be used to analyze statistical data and compute p values.

tion (2010) recommended the inclusion of effect size in the *Results* sections of research reports where the numerical results of studies are presented.[7]

Effect size (abbreviated as **ES**) is an index that is used to express the strength or magnitude of a difference between two means. It can also be used to indicate the strength of an association between two variables using correlation coefficients. Effect size is scale-free and can be used to compare outcomes from different studies where different measures are used. It is not sensitive to sample size and can be computed regardless of whether the results are statistically significant. Using effect size in addition to tests of statistical significance allows researchers to evaluate the *practical* significance of the study's finding. There are several ways to calculate effect sizes, but one of the most commonly used approaches is the index called *d*, which was developed by Cohen (1988).[8]

Effect size can be used to compare the mean scores of two groups, such as experimental and control groups, or women and men. It can also be used in experimental studies where pretest and posttest mean scores are being compared. The comparison of the means is done by converting the difference between the means of the groups into standard deviation units.

Additionally, effect size can be used in studies where the correlation between variables is calculated to assess the relationship between the variables.[9]

When interpreting statistical results, researchers should look at the direction of the outcome (for example, which mean is higher or whether the correlation is positive or negative) and whether the test statistics they compute are statistically significant. When appropriate, the effect size should also be computed to help researchers evaluate the practical importance of their data. (See chapter 9, which has an example where the effect size is used in interpreting statistical data obtained using the *t* test for independent samples.)

Once the effect size is calculated, it can then be evaluated and interpreted. While no clear-cut guidelines are available to interpret the magnitude of the effect size, many researchers follow guidelines suggested by Cohen (1988). According to Cohen, when comparing means of two groups, an effect size (that is, *d*) of 0.20 is considered small; an effect size of 0.50 is considered medium; and an effect size of 0.80 is considered large. Guidelines developed by other researchers define effect size of 0.20 as small; effect size of 0.60 as moderate; effect size of 1.20 as large; and effect size of 2.00 as very large. Effect sizes of 1.00 or higher, though, are rare in educational research.

Whether an effect size is considered practically significant may depend not only on its magnitude but also on the purpose and expectation of the researcher and the type of data being collected. For example, when assessing the efficacy of a new program that costs a great deal of money, time, and resources, an effect size of 0.60 may not be considered beneficial or cost-effective. By comparison, a school that is assessing the

---

7. HINT: See American Psychological Association. (2010). *Publication manual of the American Psychological Association* (6th ed.). Washington, DC: Author.

8. HINT: See Cohen, J. (1988). *Statistical power analysis for behavioral sciences* (2nd ed.). Hillsdale, NJ: Lawrence Erlbaum.

9. HINT: While effect size can be computed for a variety of statistical tests, we focus here on the use of effect size in studies that compare two means and in correlational studies.

effect of a new program on standardized test scores of low-ability students may view an effect size of 0.30 as very valuable if the increase in test scores allows the district to get off a list of "failing" schools.

Comparing Means

In comparing means, the index of ES is a ratio that is calculated by dividing the difference between the two means by a standard deviation. (See chapter 5 for a discussion of standard deviation.) The literature offers several approaches to obtaining the standard deviation that is used as the denominator in the equation to compute the effect size.

In experimental studies the means of experimental and control groups are usually compared in order to assess the effectiveness of the intervention. In the computation of effect size in such studies, the difference in means between the two groups is the numerator. Because the experimental group scores are usually higher than the scores of the control group, the mean of the control group is subtracted from the mean of the experimental group. The standard deviation of the control group is used as the denominator.

$$ES = \frac{\text{Mean}_{Exp} - \text{Mean}_{Cont}}{SD_{Cont}}$$

Where ES   = Effect size
     Mean_{Exp} = Mean of experimental group
     Mean_{Cont} = Mean of control group
     SD_{Cont} = Standard deviation of control group

When the two means in the numerator are two comparison groups (for example, boys and girls), the denominator is the standard deviation of both groups, combined.

An effect size around 0.00 indicates that the two groups scored about the same. A positive effect size indicates that the first mean listed in the numerator is higher than the second mean. Conversely, a negative effect size indicates that the second mean listed in the numerator was higher than the first mean.

Studying Relationship

In studying the relationship between two variables, effect sizes may be interpreted similarly to the way the correlation coefficient r is evaluated. (See chapter 7 for a discussion of correlation.) The correlation coefficient serves as an index that quantifies the relationship between two variables, and it can be used to evaluate the statistical significance, as well as the practical significance, of the study. Several researchers suggest the use of squared correlation coefficients (r^2; also known as the *coefficient of determination*) as an index of effect size in place of the correlation coefficient r. Cohen suggests the following guidelines to interpret the *practical* importance of correlation coefficients: $r = .10$ ($r^2 = .01$) is considered a small effect; $r = .30$ ($r^2 = .09$) is considered a medium effect; and $r = .50$ ($r^2 = .25$) is considered a large effect.[10] Note that while the effect size d that is used to compare means can take on values that are higher than 1.00, the correlation coefficient r can range only from -1.00 (a perfect negative correlation) to $+1.00$ (a perfect positive correlation) and r^2 cannot exceed 1.00.

USING SAMPLES TO ESTIMATE POPULATION VALUES

In research, we often want to gather data about a population that is of interest to us. As we mentioned before, in most cases it is not possible, or not practical, to study all the members of the population. Instead, we select a sample from that population and use the sample's numerical data (the sample statistics) to estimate the population values (the parameters). Keep in mind, though, that in using statistics to estimate parameters, we can expect some sampling error.

Population parameters have fixed values, but when we select a single sample from a population, the sample statistics are likely to be different from their respective population parameters. Sample values (for example, the mean) are likely to be higher or lower than the fixed population values they are designed to estimate. Nevertheless, in conducting research we usually use information from a single sample to estimate the parameters of the population from which that sample was selected. Therefore, our estimate is expressed in terms of probability, not certainty.

For example, members of a school board in a K–8 school district may be interested in investigating the attitudes of all students in the district toward school uniforms. The school board members construct a survey designed to assess students' attitudes. Instead of surveying the entire student population, the school board members select a random sample of students and administer their survey to that group of students. The responses of the students are then used to estimate the responses of all students in the district.

The school board members may ask themselves how well the responses of the sample of students represent the attitudes of the rest of the students in the district and whether the information gathered from the sample (the *sample statistics*) is an accurate representation of the population information (the *population parameters*). All of us can probably agree that a sample is not likely to be a perfect representation of the population. Therefore, when we use sample values to estimate population values, instead of studying the whole population, we risk making a certain level of error. Still, in education, as in political polls, market research, or other research studies, we usually agree to accept a certain "margin of error" and often choose to select a sample, study that sample, and use the information from the sample to make inferences about the population that is of interest to us.

---

10. HINT: Remember that squaring the correlation coefficients (r^2) yields smaller numbers than the correlation itself (r).

If we select multiple samples of the same size from the same population, compute the sample means, and plot the sample means, we would see that they are normally distributed in a bell-shaped curve. (See chapter 6 for a discussion of the normal curve.) As the number of samples increases, the shape of the distribution gets closer to a smooth-looking bell-shaped curve.

As an example, suppose we want to assess the average (that is, mean) IQ scores of the students in our high school district. If we select several samples of fifty students, and test their IQ, we can predict that there will be some variability in the mean scores obtained from each sample. The expected variation among means that are selected from the same population is considered a sampling error. Further, let's say we were to test a group of fifty students selected randomly from the population of all high school students, record their IQ scores, put their names back in the population list of names, and repeat this process over and over and select additional samples of fifty students. If we record the mean IQ of each sample, these means would form a normal distribution.

The distribution of multiple sample means of the same size that are drawn from the same population has its own mean and standard deviation. The mean indicates the location of the center of the distribution. The standard deviation (abbreviated as SD) is an index of the spread of a set of scores and their variability around the mean of the distribution. The SD indicates the average distance of scores away from the mean. (See chapters 4 and 5, respectively, for a discussion of the mean and standard deviation.) In a normal distribution, approximately 68 percent of the scores lie within plus or minus 1SD (written as ±1SD) from the mean; approximately 95 percent of the scores lie within ±2SD from the mean; and over 99 percent of the scores lie within ±3SD from the mean.

Standard Error of the Mean

The standard deviation of the sample means is called the **standard error of the mean**, and it is expressed by the symbol $SE_{\overline{x}}$. Luckily, we do not have to draw many samples in order to estimate the population mean and standard deviation. Instead, we can draw a single sample and use the information from that sample to compute the population mean and standard deviation. The mean of the sample is used to estimate the population mean and the sample standard deviation is used to estimate the population standard deviation by using this formula:

$$SE_x = \frac{SD}{\sqrt{n}}$$

Where $SE_{\overline{x}}$ = Standard error of the mean
SD = Standard deviation of the sample
n = Sample size

The standard error of the mean is used to estimate the sampling error. It shows the extent to which we can expect sample means to vary if we were to draw other samples from the same population. It can be used to estimate the sampling error we can expect if we use the information from a single sample to estimate the population standard deviation.

In the formula used to compute the standard error of the mean, the square root of the sample size is used as the denominator. Therefore, higher sample sizes would result in lower standard errors of the mean. For example, suppose we were to draw two samples from the same population—one sample with 100 members and one sample with 20 members—and compute the standard error of the mean for these two samples. We can expect the standard error of the mean of the sample with 20 members to be higher compared with the larger sample of 100 members. Our estimate of the population standard deviation would be more accurate using the standard error of the mean from the large sample because the standard error of the mean computed from the large sample would be smaller.

The standard error of the mean tells us that if we were to continue to draw additional samples of the same size from the same population, we could expect that 68 percent of the sample means would be within ±1SD of our obtained sample mean; 95 percent of the sample means would be within ±2SD of the obtained sample mean; and 99 percent of the sample means would be within ±3SD of the obtained sample mean.

Confidence Intervals

As was explained, using statistics derived from a single sample to estimate the population parameters is likely to result in a certain level of error. A **confidence interval** is a way to estimate the population value that is of interest to us. The confidence interval (CI) lets us predict, with a certain degree of confidence, where the population parameter is. A confidence interval allows us to state the boundaries of a range within which the population value we try to estimate (for example, mean) would be included in a certain percent (for example, 95 percent) of the time in samples of the same size drawn from that same population as our single sample.

The sample mean is used as the center of the confidence interval. The standard error of the mean is also used in constructing confidence intervals. The confidence interval includes two boundaries: the lower limit, represented as CI_L; and the upper limit, represented by CI_U.

Although in many situations, the goal of the researchers is to use a single sample mean to estimate the population mean, there are many other situations where researchers are interested in comparing two means from different populations. An example would be a study where experimental and control groups are compared to each other. (See chapter 9 for a discussion of the *t* test for independent samples.) Another example would be a study where researchers want to investigate the effect of an intervention by comparing pretest and posttest scores for the same group of people. (See chapter 9 for a discussion of the *t* test for paired samples.)

When we select multiple samples of the same size from two different populations, or when we repeat a certain intervention with multiple samples of the same size chosen from the same population, the *differences* between the two means are normally distributed. The distribution of the differences between the means has its own mean and standard deviation. The confidence intervals in these studies provide the lower and upper limits of the distribution of the differences between the means.

Computer programs (such as SPSS) usually provide upper and lower limits of the confidence intervals at the 95 percent confidence level. The interval's lower boundary (CI_L) and its upper boundary (CI_U) are usually reported, along with the mean and standard error of the mean. An interval of 68 percent confidence (CI_{68}) contains a narrower range compared with a confidence interval associated with 95 percent confidence. Similarly, intervals of 99 percent confidence are wider than similar intervals associated with 95 percent confidence.

The formulas used to construct confidence intervals (that are used to estimate population means in different research situations) include two statistics (*z* scores and *t* values) that are discussed later (see chapters 6 and 9). Therefore, no numerical examples are provided here to illustrate how to compute the confidence intervals and their lower and upper limits (CI_L and CI_U).

STEPS IN THE PROCESS OF HYPOTHESIS TESTING

Research studies that are conducted to test hypotheses follow similar steps. These steps are likely to be taken in studies where samples are selected and studied for the purpose of making inferences about the populations from which they were selected (that is, inferential statistics).

The hypothesis-testing process starts with the study's research question and ends with conclusions about the findings. Following is a summary of the steps in the process of the statistical hypothesis testing.

1. Formulating a research question for the study.
2. Stating a research hypothesis (that is, an alternative hypothesis). The hypothesis should represent the researcher's prediction about the outcome of the study and should be testable. Note that a null hypothesis for the study is always implied, but it is not formally stated in most cases. The null hypothesis predicts no difference between groups or means or no relationship between variables.
3. Designing a study to test the research hypothesis. The study's methodology should include plans for selecting one or more samples from the population that is of in-

terest to the researcher; selecting or designing instruments to gather the numerical data; carrying out the study's procedure (and intervention in experimental studies); and determining the statistical test(s) to be used to analyze the data. (See chapter 16 for further information about research methodology.)

4. Conducting the study and collecting numerical data.

5. Analyzing the data and calculating the appropriate test statistics (for example, Pearson correlation coefficient, t test value, or chi square value; see chapters 7, 9, and 11, respectively) and the p value.

6. Deciding whether to retain or reject the null hypothesis. A p value of .05 is the most commonly used benchmark to consider the results statistically significant. In experimental studies, if the results are statistically significant, the researcher may also wish to calculate the effect size (ES) to determine the practical significance of the study's results.

7. Making a decision whether to confirm the study's alternative hypothesis (that is, the research hypothesis) and how probable it is that the results were obtained purely by chance. This decision is based on the decision made regarding the null hypothesis.

8. Summarizing the study's conclusions, addressing the study's research question.

AND FINALLY . . .

Computer programs make the task of data analysis quick, easy, and efficient. However, it is up to the researcher to choose the appropriate statistical test, interpret the results, and evaluate their implications. For example, a difference of 3 points between two group means in one study may not be as meaningful as the same finding in another study.

Statistical analyses are based on observations that are collected using certain instruments and procedures. If the instruments used to collect data lack in reliability or validity, any conclusions or generalizations based on the results obtained using these instruments are going to be questionable. Similarly, when a study is not well designed, one may question the results obtained. Problems resulting from a poorly designed study and bad data cannot be overcome with a fancy statistical analysis. Just because the computer processes the numbers and comes up with "an answer" does not mean that these numbers have any real meaning. Remember what is often said regarding the use of computers: "Garbage in, garbage out." This adage applies to the use of statistics as well.

SUMMARY

1. The term **statistics** refers to methods and techniques used for describing, organizing, analyzing, and interpreting numerical data.

2. A **variable** is a measured characteristic that can vary and assume different values or levels.

3. A **constant** is a measured characteristic that has only one value.

4. Variables may be *continuous* or *discrete*. **Continuous variables** can take on a wide range of values and contain an infinite number of small increments. **Discrete variables** contain a finite number of distinct values between two given points.

5. **Measurement** is defined as assigning numbers to observations according to certain rules. There are four types of measurement scales: *nominal, ordinal, interval*, and *ratio*.

6. In **nominal scales**, numbers are used to label, classify, or categorize observations to indicate similarities or differences. This is the least precise form of measurement.

7. In **ordinal scales**, observations are ordered to indicate *more than* or *less than* based on magnitude or size. The intervals between the observations, though, cannot be assumed to be equal.

8. In **interval scales**, observations are ordered with equal intervals between points on the scale. Since there is no absolute zero point, inferences cannot be made that involve ratio comparisons.

9. In **ratio scales**, observations are ordered with equal intervals between points. This scale has an absolute zero; therefore, comparisons can be made involving ratios. This is the most precise form of measurement. Ratio scales are generally used in physical sciences rather than in the behavioral sciences.

10. A **population** is the entire group of persons or things that have some characteristic in common.

11. A **sample** is a group of observations or cases selected from the total population.

12. Some chance variation in sample numerical values (for example, mean) occurs when we repeatedly select same-size samples from the same population and compare their numerical values. This error, called a *sampling error*, is beyond the control of the researcher.

13. A **parameter** is a measure of a characteristic of the entire population.

14. A **statistic** is a measure of a characteristic of the sample.

15. A sample should be *representative* of the population because the statistics gained from the sample are used to estimate the population parameters.

16. Most research studies in education are designed to study populations by using samples that are representative of these populations.

17. In selecting a **simple random sample**, every member of the population has an equal and independent chance of being selected.

18. In **systematic sampling**, every *Kth* person is selected from the population. *K* is determined by dividing the total number of population members by the desired sample size.

19. The first step in obtaining a **stratified sample** is to divide the population into subgroups (called *strata*), followed by a random selection of members from each subgroup. The final sample represents, proportionately, the various subgroups in the population.

20. A **convenience** (or **incidental**) **sample** is a sample that is readily available to the researcher. Researchers have to exercise great caution in generalizing results from a convenience sample to the population.

21. **Sample bias** refers to *systematic* (as opposed to *random*) differences between the sample and the population from which it was selected. A biased sample contains a certain systematic error.

22. As sample size increases, it is more likely to be an accurate representation of the population, especially when the sample is randomly chosen. In many research studies, a sample size of at least thirty is desirable. However, size alone does not guarantee that the sample is representative, and a large sample may still be biased.

23. There is a clear relationship between the sample size and the confidence level in rejecting the study's null hypothesis. As the sample size *increases*, a *lower* computed test statistic value is needed in order to reject the null hypothesis at the $p = .05$ level.

24. In well-designed experimental studies, a sample that is a smaller portion of the population may be sufficient to provide an accurate representation when chosen from larger population sizes.

25. **Parametric statistics** (also called *assumption-free statistics*) are applied to populations that meet certain requirements. **Nonparametric statistics** can be applied to all populations, even those that do not meet the basic assumptions.

26. Parametric statistics are used more often by researchers and are considered more powerful and more efficient than nonparametric statistics.

27. Nonparametric statistics can be used with nominal, ordinal, interval, and ratio scales, whereas parametric statistics can be used with interval and ratio scales only.

28. **Descriptive statistics** classify, organize, and summarize numerical data about a particular group of observations.

29. **Inferential statistics** (also called *sampling statistics*) involve selecting a sample from a defined population, studying the sample, and using the information gained to make inferences and generalizations about that population.

30. Descriptive and inferential statistics are not mutually exclusive, and the same measures can be used in both types. The purpose, or the use of the statistics, determines whether they are descriptive or inferential.

31. A **hypothesis** is a prediction (an "educated guess") about the outcome of the study. After the hypothesis is proposed, a study is designed to test the hypothesis.

32. The main hypothesis proposed by the researcher about the study's outcome is called the **alternative hypothesis** (or simply *the hypothesis*). It is represented by the symbol H_A or H_1.

33. A **null hypothesis** (also called a *statistical hypothesis*) always states that there would be no differences between groups or means being compared, or no relationship between variables being correlated beyond what might be expected purely by chance. A null hypothesis is represented by the symbol H_0.

34. Hypotheses may be stated as *directional* or *nondirectional*. A **directional hypothesis** predicts the direction of the outcome of the study. A **nondirectional hypothesis** predicts that there will be a statistically significant difference or relationship, but the direction is not stated.

35. In studies where *differences* are investigated, a directional hypothesis predicts which group will score higher on the dependent variable. A nondirectional hypothesis predicts a difference in scores on the dependent variable, but not the direction of the difference (that is, which group's mean would be higher).

36. In studies of *relationship*, a directional hypothesis predicts whether the relationship (for example, correlation) would be positive or negative. A nondirectional hypothesis predicts that the variables would be related, but it does not specify whether the relationship would be positive or negative. A null hypothesis predicts no relationship between the variables.

37. The process of the statistical hypothesis testing starts with a decision regarding the null hypothesis.

38. The study's statistical results are used to decide whether the null hypothesis should be retained or rejected. In studies where the alternative hypothesis is directional or nondirectional, *rejecting* the null hypothesis usually leads to the confirmation of the alternative hypothesis, while *retaining* the null hypothesis usually leads to a decision not to confirm the alternative hypothesis.

39. Results can be reported as **statistically significant** or *not statistically significant*. When the results are statistically significant, the *exact* level of significance may be reported.

40. Statistical results are reported in terms of *probability*, not *certainty*. Results that are statistically significant are usually reported in terms of probability (**p value**), or level of significance, using terms such as $p < .05$ or $p < .01$.

41. When the probability level is set at the beginning of the study, *before* collecting and analyzing the data, it is represented by the Greek letter *alpha* (α). The convention is to use an alpha level of .05.

42. A **Type I error** is made when researchers decide to *reject* the null hypothesis (H_o) when in fact it is true and *should not be rejected*.

43. A **Type II error** is made when researchers decide to *retain* the null hypothesis, when in fact it *should be rejected*.

44. The proper decision is made when researchers reject a false null hypothesis, or when they retain a true null hypothesis.

45. The **degrees of freedom** (**df**) in most statistical tests are calculated as $n - 1$ (the number of people in the study, minus 1). Researchers routinely get *df* (as well as *p* values) from computer statistical programs.

46. In interpreting statistical results, researchers should look at the *direction* of the outcome and whether the results are *statistically significant*. When appropriate, *effect size* should be computed to evaluate the *practical significance* of the data.

47. **Effect size (ES)** is an index that is used to express the strength or magnitude of difference between two means or the relationship between variables. Using

the index of effect size allows researchers to evaluate the *practical significance* in addition to the *statistical significance* of their studies.

48. There are several ways to calculate effect sizes. One of the most commonly used effect sizes, called *d*, was developed by Cohen (1988). ES is calculated by dividing the difference between the two means by a standard deviation (SD).

$$ ES = \frac{\text{Mean}_1 - \text{Mean}_2}{\text{SD}} $$

49. While no clear-cut guidelines are available for the interpretation of effect size, many researchers follow those offered by Cohen in 1988: *ES* of 0.20 is considered small; *ES* of 0.50 is considered medium effect; and *ES* of 0.80 is considered large.

50. The effect size may be interpreted similarly to the way the correlation coefficient *r* is evaluated. Researchers use *r* or r^2 as an index of effect size. Cohen suggested the following guidelines for evaluating correlation coefficients: $r = .10$ ($r^2 = .01$) is considered small; $r = .30$ ($r^2 = .09$) is considered medium; and $r = .50$ ($r^2 = .25$) is considered large.

51. Population parameters have fixed values, but when we select a single sample from a population, the sample statistics are likely to be different from the respective population parameters. Sample values (for example, means) are likely to be higher or lower than the fixed population values they are designed to estimate.

52. If we select multiple samples of the same size from the same population, compute the sample means, and plot the sample means, we would see that they are normally distributed in a bell-shaped curve.

53. The distribution of multiple sample means of the same size that are drawn from the same population has its own mean and standard deviation.

54. The standard deviation of the sample means is called the **standard error of the mean**. A single sample can be used to compute the standard error of the mean, which can be used as an estimate of the standard deviation of the population. The sample standard deviation and size are used in the formula to compute the standard error of the mean:

$$ SE_x = \frac{\text{SD}}{\sqrt{n}} $$

55. The index of the standard error of the mean is used to estimate sampling error. It shows the extent to which we can expect sample means to vary if we were to draw other samples from the same population. It can be used to estimate the sampling error we can expect if we use the information from a single sample to estimate the population standard deviation.

56. The standard error of the mean tells us that if we were to continue to draw additional samples of the same size from the same population, we could expect that 68 percent of the sample means would be within ±1SD of our obtained sample mean; 95 percent of the sample means would be within ±2SD of the obtained sample mean; and 99 percent of the sample means would be within ±3SD of the obtained sample mean.

57. The **confidence interval** (**CI**) allows us to state the boundaries of a range within which the population value (for example, the mean) we try to estimate would be included. The interval lets us predict, with a certain degree of confidence, where the population parameter is expected to be.

58. The sample mean serves as the center of the interval used to estimate the population mean. The standard error of the mean is also used in constructing confidence intervals.

59. The confidence interval includes two boundaries: the *lower limit*, represented as CI_L; and the *upper limit*, represented by CI_U.

60. When we select multiple samples of the same size from two different populations, or when we repeat a certain intervention with multiple samples of the same size chosen from the same population, the *differences* between the two means are normally distributed. This distribution has its own mean and standard deviation. The confidence intervals in these studies provide the lower and upper limits of the distribution of the differences between the means.

61. The confidence level of 95 percent (CL_{95}) is used the most and it is the one that is reported most often on the printouts of statistical software programs (for example, SPSS). The interval's lower boundary (CI_L) and its upper boundary (CI_U) are usually reported, along with the mean and standard error of the mean.

62. Research studies that are conducted to test hypotheses follow similar steps. The hypothesis-testing process starts with the study's research question and ends with conclusions about the findings. These steps are likely to be taken in studies where samples are studied for the purpose of making inferences about the populations from which they were selected.

CHECK YOUR UNDERSTANDING

1. Suggest examples of data measured on *nominal, ordinal, interval,* and *ratio* scales.
2. Provide examples of studies where researchers might select *random, systematic, stratified,* and *convenience* samples.
3. Describe a study that you may conduct in your setting or in other settings where you will investigate a *sample*:
   a. Explain what type of sample you will use and how you will select it.
   b. Propose a hypothesis for the study; this hypothesis can be *directional* or *nondirectional*.
   c. What would be the *null* hypothesis for your study?
   d. How would you know whether your hypothesis was supported? Explain.
4. What is the difference between *statistical* and *practical* significance? Give examples and provide explanations to illustrate the difference.

II

DESCRIBING DISTRIBUTIONS

Organizing and Graphing Data

Chapter 3 focuses on ways to *organize* numerical data and present them visually with the use of *graphs*. Data that you have access to or have collected as part of your study often do not appear well organized, and it is hard to see trends or compare results from one set of numbers to another. Organizing your information helps you get a better "feel" for your data. One way to organize your data is to create a frequency distribution. Next, when you are ready to graph your data, you have several types of graphs to choose from. Choices include *histogram* and *frequency polygon*, *pie graph*, *bar graph*, *line graph*, and *box plot*. Explanations about the use of each shape, guidelines for its appropriate use and how to construct it, and specific examples of each type of graph are provided throughout the chapter.

Various software programs, such as Excel, can easily produce graphs for you. Remember, though, that it is your responsibility to choose the right graphs for displaying your data and to produce accurate and clear graphs that will present your data visually and accurately and be easy to read and interpret.

Data collected in education or in other fields often come as a series of numbers. An example may be a list of test scores from a weekly quiz administered by classroom teachers. In order to better understand and make use of the data, there is a need to organize and summarize the scores. Numerical and graphical presentations of the data are helpful in making an easier and more efficient use of these data. In this chapter, we introduce you to several ways you can organize and illustrate your data visually.

ORGANIZING DATA

Frequency Distributions

Organizing and graphing data allows researchers and educators to describe, summarize, and report their results. By organizing data they can compare distributions and observe patterns. In most cases, though, the original information we collect is not ordered or summarized. Therefore, after collecting data, we may want to create a **frequency distribution** by ordering and tallying the scores.

To illustrate the use of frequency distributions, let's look at the following example. A seventh grade social studies teacher wants to assign end-of-term letter grades to the twenty-five students in her class. After administering a 30-item final examination, the teacher records the students' test scores next to each student's name (table 3.1). These scores show the number of correct answers obtained by each student on the exam. Next, the teacher can create a frequency distribution by ordering and tallying these test scores (table 3.2).

Table 3.1.   Scores of Twenty-Five Students on a 30-Item Social Studies Test

27	16	23	22	21
25	28	26	20	22
30	24	29	17	26
24	17	23	24	25
19	21	18	23	25

Table 3.2.   A Frequency Distribution of the Scores Presented in Table 3.1 Ordered from the Highest to the Lowest Score

Score	Frequency	Score	Frequency
30	1	22	2
29	1	21	2
28	2	20	1
27	1	19	1
26	2	18	1
25	2	17	2
24	3	16	1
23	3		

Note that the list of scores is still quite long; the highest is 30 and the lowest is 16. The teacher may want to group every few scores together in order to assign letter grades to the students. The following is a discussion of the process used for grouping scores.

Class Intervals

When the ordered list of scores in the frequency distribution is still quite long, as is the case in our example (see table 3.2), the teacher can group every few scores together into class intervals. **Class intervals** are usually created when the range of scores is at least 20. The recommended *number* of intervals should be eight to twenty.[1] The biggest disadvantage of using class intervals is that we lose details and precision. That is, because scores are grouped, we cannot tell what the *exact* score obtained by each person was. For example, assume we know that there are four students in a class interval of 20-25. We cannot tell which score was obtained by each of these students, only that their scores were between 20 and 25.

Two rules should be observed when creating class intervals: (a) all class intervals should have the same width; and (b) all intervals should be mutually exclusive (that is, a score may not appear in more than one interval). Whenever possible, the width of the interval should be an odd number to allow the midpoints of the intervals to be whole numbers.

Table 3.3 contains test scores of thirty students on an 80-point test. The lowest score obtained by a student on the test is 31, and the highest score is 80. The scores in the table are listed in descending order. Table 3.4 shows the same scores, grouped into ten class intervals. Each interval has a width of 5 points.

Table 3.3. Test Scores of Thirty Students on an 80-Point Test

80	62	57	52	44
74	61	57	51	43
69	59	56	50	41
66	58	55	49	39
65	58	54	48	36
63	57	53	47	31

Table 3.4. A Frequency Distribution of the Thirty Scores in Table 3.3 with Class Intervals of 5 Points and Interval Midpoints

Class Intervals (5 Points)	Interval Midpoint	Frequency
76–80	78	1
71–75	73	1
66–70	68	2
61–65	63	4
56–60	58	7
51–55	53	5
46–50	48	4
41–45	43	3
36–40	38	2
31–35	33	1

---

1. HINT: There are no strict rules as to the number of class intervals that should be used. Some textbooks recommend eight to twenty intervals, while others recommend ten to twenty. In the examples in this chapter we used ten intervals (table 3.4) and eight intervals (table 3.8). Most computer statistical programs are likely to create class intervals for you, so the computational steps in the book are used mostly to explain the concept of class intervals.

Cumulative Frequency Distributions

Another way to organize data is to create a *cumulative frequency distribution*. A **cumulative frequency distribution** shows the number of scores *at* or *below* a given score. Percentages are often added to these tables. (Computer programs such as SPSS routinely generate cumulative frequency tables.) Table 3.5 is a cumulative frequency table showing the test scores of twenty students. The table starts with a frequency distribution in column 1 and column 2 (similar to the frequency distribution in table 3.2). These two columns are titled *Score* and *Frequency*.[2] In the third column, *Percent Frequency*, the frequencies that are listed in the second column are converted into percentages. For example, inspecting the top of column 2, we can see that one student had a score of 20. Because there are twenty students in the class, we can say that 5 percent (1 out of 20) of those students had a score of 20. Similarly, we can see that 10 percent (2 students out of 20) had a score of 17.

In the fourth column, *Cumulative Frequency*, entries are created from the bottom up. For example, to calculate the first entry at the bottom of column 4, look at the lowest numbers in columns 1 and 2 and ask yourself the following: How many students had a *score of 5* (the lowest score in this distribution) or *less*? The answer is "1"; therefore, "1" is recorded at the bottom of column 4.

Next, ask yourself: How many students had a *score of 6* or *less*? (To find the answer to this question, add up the two lowest numbers in column 2.) The answer is "3," and this number is recorded right above "1" at the bottom of the fourth column. Now, calculate the number of students who had a *score of 8* or *less*. The answer is "5," the third lowest number in column 4. Continue constructing column 4 all the way up until the column has been completed.

Table 3.5. **Cumulative Frequencies Distributions of Test Scores of Twenty Students**

(Col. 1) Score	(Col. 2) Frequency	(Col. 3) Percent Frequency	(Col. 4) Cumulative Frequency	(Col. 5) Cumulative Percentage
20	1	5	20	100
19	1	5	19	95
17	2	10	18	90
16	4	20	16	80
14	4	20	12	60
10	3	15	8	40
8	2	10	5	25
6	2	10	3	15
5	1	5	1	5
	N=20			

The fifth column is titled *Cumulative Percentage*. To create this column, the cumulative frequencies in column 4 are converted to percentages. The conversion

---

2. A HINT: Note that the number of scores in this distribution is indicated at the bottom of the second column as "N = 20." In statistics, the symbol N or n is used to represent the number of cases or scores.

can be done by starting either at the top or at the bottom of column 5. To compute each entry in column 5, convert the corresponding cumulative frequency in column 4 into a percentage. To do so, divide the number in column 4 by 20, the total number of students in our example. If you work your way from the top down, the first cumulative frequency you need to convert to a percentage is 20 (the top number in column 4). In other words, you have to calculate what percentage of students in the class had a score of 20 or less. Clearly, *all* 20 students in the class had a score of 20 or less. Therefore, enter 100 (that is, 100 percent) at the *top* of the fifth column. Next, convert 19 (the second number in column 4) into percentages. Nineteen out of 20 (the total number of students in the class) is 95 percent, which appears as the second number in column 5. Continue to work all the way down to the bottom of column 5 until it is complete.

Following the same steps as those used to create the entries in table 3.5, classroom teachers can use scores from tests they give their students to compute the students' cumulative percentiles. These cumulative percentiles are also called **percentile ranks**, and they are used to compare the performance of students in the class to that of their classmates. (See chapter 6 for a more comprehensive discussion of percentile ranks.) Using the data in table 3.5 as an example, we can say that a student with a score of 19 had a percentile rank of 95. This percentile rank means that the student did better than, or as well as, 95 percent of the other students in the class who took the examination at the same time as that student. Similarly, a student with a score of 14 had a percentile rank of 60 and did better than, or as well as, 60 percent of the students in the class.

GRAPHING DATA

Graphs are used to communicate information by transforming numerical data into a visual form. Graphs allow us to see relationships not easily apparent by looking at the numerical data. There are various forms of graphs, each one appropriate for a different type of data. While many computer software programs provide a dazzling array of graphic choices, it is the responsibility of those creating the graphs to select the right graph for their data. The rest of this chapter discusses various graphs and how they can be used.

Histogram and Frequency Polygon

Frequency distributions, such as the one in table 3.6, can be depicted using two types of graphs, a **histogram** (figure 3.1, *part a*) or a **frequency polygon** (figure 3.1, *part b*).

Table 3.6.   A Frequency
Distribution of Thirteen Scores

Score	Frequency
6	1
5	2
4	4
3	3
2	2
1	1

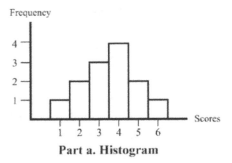

Part a. Histogram

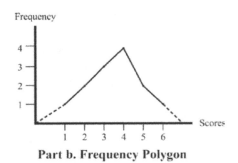

Part b. Frequency Polygon

FIGURE 3.1
Histogram and frequency polygon of the data in table 3.6

In drawing histograms and frequency polygons, the vertical axis *always* represents frequencies, and the horizontal axis *always* represents scores or class intervals. The lower values of both vertical and horizontal axes are recorded at the intersection of the axes (at the bottom left side). The values on both axes increase as they get farther away from the intersection (figure 3.2).

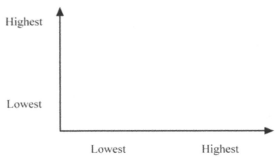

FIGURE 3.2
Graphing the lowest and highest values on each axis

Both the histogram and the frequency polygon can be used to graph individual scores as well as scores grouped into class intervals. When graphing class intervals, tick marks on the horizontal axis show the interval midpoint. The upper and lower scores in each interval may also be recorded instead of the interval midpoint. Table 3.7 shows English test scores of twenty-five students, and figure 3.3 presents the same data using a frequency polygon.

Table 3.7. A Frequency Distribution of Twenty-Five Scores with Class Intervals and Midpoints

Class Interval	Midpoint	Frequency
38–42	40	1
33–37	35	3
28–32	30	4
23–27	25	6
18–22	20	5
13–17	15	3
8–12	10	2
3–7	5	1

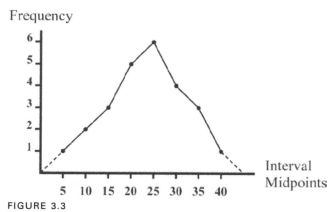

FIGURE 3.3
A frequency distribution with interval midpoints representing class intervals of the data in table 3.7

As the number of scores in the distribution increases, it is more likely to have smoother curve lines, compared with a distribution with fewer scores. Distributions with a small number of cases are likely to have curve lines that are more jagged and uneven. The bell-shaped normal distribution that all of us are familiar with is actually a special case of a frequency polygon with a large number of cases. (See chapter 6 for a discussion of the normal curve and bell-shaped distributions.)

Comparing Histograms and Frequency Polygons

To some extent, the decision about whether to use a histogram or a frequency polygon is a question of personal choice. One advantage of the polygon over the histogram is that the polygon can be used to compare two groups by displaying both groups on the same graph. For example, figure 3.4 shows the test scores of a large group of boys and a large group of girls on a biology test. As can be easily seen, the ranges of both groups were about the same. However, the girls overall performed better than the boys and obtained higher scores.

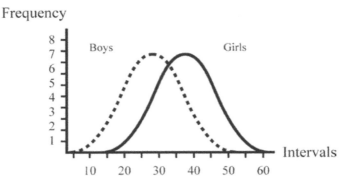

FIGURE 3.4
A frequency polygon comparing scores of boys and girls

Several textbooks list another difference between a histogram and a frequency polygon. These books suggest using histograms for data that naturally occur in discrete units. For example, people's IQ scores may be reported to be 101, 102, or 103, but not 101.5 or 102.5.[3] By comparison, frequency polygons are recommended for depicting distributions of scores that can take on in-between values, such as height and weight. For example, let's say we want to use a frequency polygon to graph the heights of the students in our class, rounded off to the nearest inch. Using 1-inch increments to mark the horizontal axis, the frequency polygon would show only data points such as 5'1", 5'2", or 5'3" even though in reality we may have students whose height is somewhere between 5'1" and 5'2", or between 5'2" and 5'3". It is important to remember that although the scores on the variable being graphed are measured on a continuum, the frequency polygon itself may show only discrete units.

Pie Graph

The **pie graph** (or **pie chart**) looks like a circle that is divided into "wedges," or "segments." Each wedge represents a category or subgroup within that distribution. The size of each wedge indicates the percent of cases represented by that wedge. By inspecting the pie graph, we can readily see the proportion of each wedge in relation to the total pie as well as the relationships among the different wedges. The percentages represented by the different wedges should add up to 100 percent.

When drawing a pie graph, the different wedges of the pie should be identified and numerical information, such as percentages, should be included. This would allow

---

3. HINT: Although IQ scores for *individuals* are reported only as whole numbers, *group* mean scores may have decimal places, such as 103.4 or 104.8.

easy and accurate interpretation of the graph. There should not be too many wedges in the pie circle. The fifth edition of the *Publication Manual of the American Psychological Association* recommended that the pie graph be used to compare no more than five items and that the wedges be ordered from the largest to the smallest, starting at 12 o'clock.[4] However, it is not uncommon to see reports that include pie graphs with more than five wedges.

To illustrate how to draw and interpret pie graphs, study the data in table 3.8 and the graph in figure 3.5. The table and pie graph show the proportions of four feeder elementary schools in the total population of a junior high school. Note that the total percentages of the four schools add up to 100 percent.

Table 3.8. Proportions of Four Feeder Elementary Schools in a District Junior High School

School	Percent in Jr. High
Grand Oak	20
Jefferson Elementary	35
Ocean View	35
Sunset Hill	10
TOTAL	100

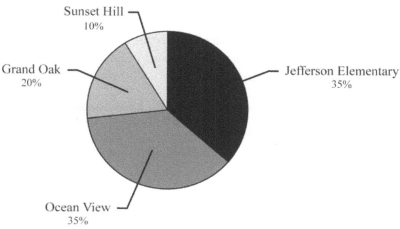

FIGURE 3.5
A pie graph showing the data in table 3.8

Pie graphs lend themselves well to making comparisons by placing two pie graphs next to each other. For example, we can create two side-by-side pies to show changes in demographic characteristics (for example, racial or ethnic groups) in a city. One pie graph can depict the demographic data from one year, and the other graph can show

---

4. HINT: The complete reference for the previous edition of the APA publication manual is: American Psychological Association. (2001). *Publication manual of the American Psychological Association* (5th ed.). Washington, DC: Author.

the same demographic data from another year. This would allow those studying the two graphs to easily see changes from one year to the next. Or the two pies can represent demographic data from two different cities to allow a comparison of the two. Another popular way to use two pies side by side is to use one of the pies to depict income and the other pie to depict expenses in presenting a budget to some constituency. By looking at the two pies, it is easy to see what the main sources of income are and what the main categories of expenses are. To demonstrate that the budget is balanced, the total dollar amount in the pie showing expenses should not exceed the dollar amount in the pie showing the income.

Bar Graph

A **bar graph** (also called a **bar diagram**) is a graph with a series of bars next to each other. The bars represent *discrete* categories, and they are ordered in some way, usually from the highest to the lowest or from the lowest to the highest. Although the bars are placed next to one another, they *should not touch* each other.

To illustrate the use of a bar graph, let's say we want to compare the level of support for school uniforms for students among four groups: administrators, parents, students, and teachers. Five hundred respondents from each group indicated whether they support school uniforms, and the results are recorded in table 3.9. The actual number of respondents in each group who support school uniforms was converted into percentages and recorded in the table. Inspecting the data in table 3.9 makes it clear that there are differences among the opinions of administrators, parents, students, and teachers regarding school uniforms.

Table 3.9. Comparing the Responses of Administrators, Parents, Students, and Teachers to the Question: "Do You Support School Uniforms?"

Group	% Supporting School Uniforms
Administrators	85
Parents	68
Students	30
Teachers	77

Note that the four groups in table 3.9 comprise categories that are independent of each other. In other words, the responses of people in one group do not affect or change the responses of people in another. Note also that the percentages recorded in the column on the right side do not add up to 100 percent because each row represents an independent category (a group of respondents). This is different from pie graphs, where the percentages add up to 100 percent.

Before drawing the bar graph in figure 3.6, we need to decide how to order the bars. Although the groups are listed in alphabetical order in the "Group" column, we need to reorder the groups so that the bar graph we draw has bars that are ordered by height. Therefore, we can start with the bar representing the administrators, the group with the highest percentage (85%). This bar is followed by the bars representing the other three groups: teachers (77%), parents (68%), and students (30%).

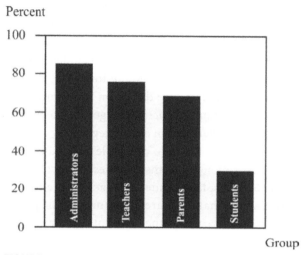

FIGURE 3.6
A bar graph showing the data in table 3.9

The bars may also be drawn horizontally to allow for easier identification and labeling of each bar. Figure 3.7 shows the same data as in figure 3.6.

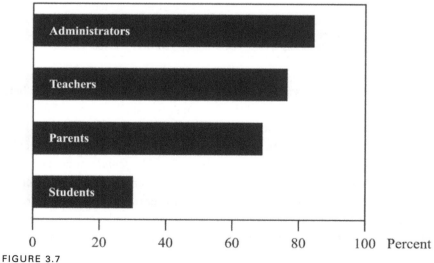

FIGURE 3.7
A horizontal bar graph showing the same data as in figure 3.6

Bar graphs may look a bit like histograms because both of them have a series of bars next to each other. However, they are used for different purposes. The histogram, as you remember, is used to display frequency distributions. In a histogram, the horizontal bar shows scores or class intervals and the vertical axis shows frequencies. The scores (or class intervals) in a histogram comprise an interval or a ratio scale, and they are on a continuum in numerical order. By comparison, the bars in a bar graph represent nominal, categorical data (for example, several groups of respondents), and *do not* imply a continuum. To indicate that each bar is independent of the other bars and represents discrete data, these bars should not touch.

Occasionally, we may want to compare data from two or more groups *within* each category, in addition to comparing data *across* categories. To facilitate such comparisons, we can use multiple *joint bars* within each category.

Let's say we want to compare the ratio of female and male students who are enrolled in different college major programs: Business, Education, Engineering, and Social Sciences. Table 3.10 displays gender data across the four majors. After inspecting the table, we can see that there are differences in the proportions of female and male students enrolled in each of the four college majors. An overwhelming majority of students in education are female (80%), and the majority of students in engineering are male (83%), while the majority of students in Social Sciences are female (64%). The number of female and male students in Business is practically even.

Table 3.10. Enrollment Breakdown by Gender in Four Undergraduate College Majors (in Percent)

Major	Females	Males	TOTAL
Business	49	51	100
Education	80	20	100
Engineering	17	83	100
Social Sciences	64	36	100

The joint bars in figure 3.8 represent the data in table 3.10. The differences between the lengths of the joint bars in each of the four majors visually express the differences in gender enrollment in these majors.

When drawing a bar graph with joint bars, it may be difficult at times to decide how to order the bars. In figure 3.8, the joint bars are ordered by *female* student enrollment.

Therefore, the first set of joint bars shows the gender enrollment figures representing *Education*, and the last set of joint bars shows the enrollment figures representing *Engineering*. If we had decided to order the bars by *male* student enrollment figures, then the joint bars representing *Engineering* should be first and the joint bars representing *Education* enrollment should be last.

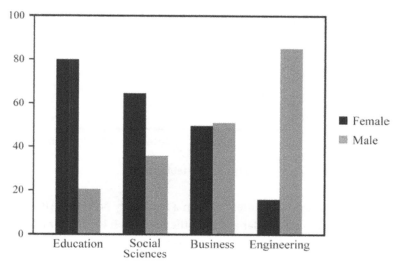

FIGURE 3.8
A bar graph with joint bars showing the data in table 3.10

Line Graph

A **line graph** is used to show relationships between two variables, which are depicted on the two axes. The horizontal axis indicates values that are on a continuum (for example, calendar years or months). The vertical axis can be used for various types of data (for example, test scores, temperatures, and amount of income). A line connects the data points on the graphs. Table 3.11 shows mean test scores of second grade students in one school over the last four years, and figure 3.9 shows the same information graphically.

Table 3.11.   Mean Test Scores of Second
Graders, 2012–2015

Year	Mean Test Score
2012	56
2013	74
2014	66
2015	81

A big advantage of the line graph is that more than one group can be shown on the same graph simultaneously. If you cannot use colors, each group can be presented by a different kind of line (for example, broken or solid). Figure 3.10 shows mean test scores of two groups of students over a four-year period.

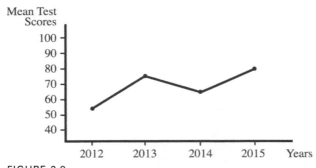

FIGURE 3.9
A line graph showing the data in table 3.11

Notice that the line graph is different from a frequency polygon. Although the two graphs may look somewhat similar at times, the line graph is used for a different purpose and does not display frequencies on the vertical axis.

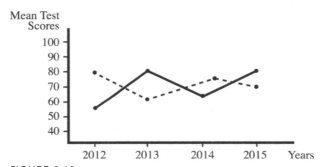

FIGURE 3.10
A line graph showing mean test scores of two groups over a four-year period

Box Plot

The **box plot** (which is also called **box-and-whiskers**), consists of a box and two whiskers and is used to depict the spread and skewness of frequency distributions. This graph was developed by John Tukey in 1977. The box represents the middle 50 percent of the distribution, and the whiskers represent the top and bottom 25 percent. A horizontal line inside the box represents the median of the distribution. (See chapter 4 for a discussion of the median.) The lengths of the whiskers indicate the spread of the scores.

To create the box plot, the scores are first ordered and divided into four quartiles, identified as Q_1, Q_2, Q_3, and Q_4. The two middle quartiles (Q_2 and Q_3) are located *within* the box, whereas the two extreme quartiles (Q_1 and Q_4) are displayed using vertical lines (the whiskers) *outside* the box.

To illustrate how to use, construct, and interpret a box plot, let's look at the following example. An eighth grade mathematics teacher wants to examine different ways to practice estimation with her class of twenty students. Each student is asked to estimate, to the nearest inch, the length of a 30-inch stick. The students' estimates are recorded in table 3.12 in the "Prepractice" column. Next, the teacher practices with the students several different ways of estimating length and each student is asked to estimate the length of the stick again. These estimates are recorded as "Postpractice" in the table.

Table 3.12. Prepractice and Postpractice Scores for Twenty Eighth Grade Students

Student Number	Prepractice	Postpractice
1	13	24
2	18	25
3	22	25
4	24	26
5	27	27
6	28	28
7	28	28
8	29	29
9	29	29
10	30	30
11	32	31
12	32	31
13	33	32
14	35	32
15	38	32
16	39	33
17	39	34
18	42	34
19	45	35
20	49	36

To display the changes in the students' prepractice and postpractice estimation scores, the teacher creates a box plot graph (see figure 3.11). Notice that the median scores of the prepractice and postpractice are similar (31.60 and 30.50, respectively). However, the range of the scores on the prepractice estimation is much higher than the range on the postpractice estimation. The scores on the prepractice range from 13 to 49 compared with a range of 24 to 36 on the postpractice. Note also that the whiskers of the prepractice scores are much longer, indicating a greater range of scores in the top and bottom quartiles, compared with the postpractice whiskers, which are quite short. Further, the two middle quartiles (Q_2 and Q_3), which include the middle 50 percent of the scores (the "boxes"), are narrower on the postpractice compared with the prepractice.

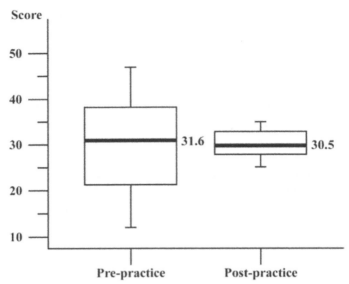

FIGURE 3.11
A box plot of the prepractice and postpractice scores showing the
data in table 3.12

Based on the data in table 3.12 and the box plot in figure 3.11, the mathematics
teacher observes the following: (a) the most extreme prepractice scores were elimi-
nated, (b) the range of scores decreased from prepractice to postpractice, and (c) the
whiskers became shorter on the postpractice. The teacher concludes that the class
practice exercises improved the students' ability to estimate the length of the stick.

DRAWING ACCURATE GRAPHS

This chapter describes some of the most popular and commonly used graphs, as well
as some of the rules and guidelines for drawing such graphs. Because most of us depend on statistical software packages to generate graphs for us, we should carefully
check the different options available on the software packages we use. Graphs should
offer an accurate visual image of the data they represent and be easy to read and interpret. Each graph should be clearly labeled, and different parts of the graph should
be identified. Graphs that are too "dense" or include too much information can be
confusing and hard to follow. This is especially true for graphs in printed materials
where colors are not available (for example, hard copies of journals). For example, pie
graphs with too many wedges or line graphs that depict too many groups may become
too "dense" and difficult to read.

To illustrate this point, look at the line graph in figure 3.12. This graph shows mean
test scores of six groups of students over four years. Inspecting the graph, we can see
that having so many lines in one graph makes it difficult to compare the groups or
observe trends over time.

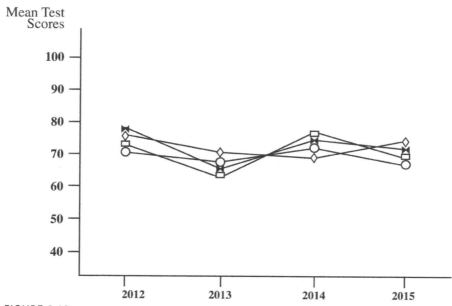

FIGURE 3.12
Line graph showing many horizontal lines that create a confusing graph

The *scale* one chooses for drawing graphs can affect the message conveyed by the graph. For example, let's look at figure 3.13 and study the two graphs in *part a* and *part b*. This figure shows the graduation rates (in percentages) of students in four high school districts. Note that while the two graphs look similar, the vertical axis in *part a* starts with a score of 0 (zero) while the vertical axis in *part b* starts with a score of 50. Consequently, the differences among the four bars that represent the districts in *part b* look much more pronounced than in *part a*. While the scale of the bars in *part b* shows more details, it also magnifies the differences and presents an exaggerated picture of them.

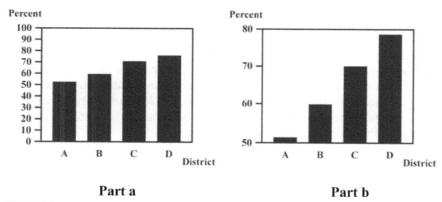

Part a   **Part b**

FIGURE 3.13
Graduation rates in four high school districts: A comparison of two graphs showing the same data

SUMMARY

1. The first step in organizing data is to order the scores in some manner. For example, the scores can be ordered from highest to lowest or from lowest to highest.
2. When scores in a distribution repeat more than once, researchers may choose to tally the scores in order to eliminate the need to list duplicate scores.
3. A **frequency distribution** is a list of scores that have been ordered and tallied.
4. When a distribution has a wide range of scores (a range of twenty or more in most cases), it is recommended that every few scores in that distribution be grouped into **class intervals**. However, when class intervals are used, detail and precision are lost and there is no way to know what the exact scores in each interval are.
5. When using class intervals, all intervals should have the same width and no score should appear in more than one interval.
6. Data may be organized using a **cumulative frequency distribution**, which shows the number and percentage of scores at or below a given score.
7. **Graphs** convey numerical information in a visual form. It is important to choose the proper graph for each type of data.
8. **Histograms** and **frequency polygons** are used to visually display frequency distributions.
9. When drawing either a histogram or frequency polygon, the following guide-lines should be used: (a) the vertical axis represents frequencies and the horizontal axis represents scores or class intervals, and (b) the lower values of both axes are at the point of their intersection. (Other guidelines for draw-ing histograms and frequency polygons are also discussed in the chapter.)
10. When drawing histograms or frequency polygons to show grouped data, midpoints or exact scores are often used to mark each class interval.
11. The **pie graph** (or **pie chart**) looks like a circle that is divided into "wedges," or "segments." Pie graphs allow us to see relationships among the different wedges that comprise the total distribution. The size of each wedge indicates the proportion of cases in that wedge.
12. The **bar graph** (or **bar diagram**) is used to represent nominal, categorical data. It has a series of bars that do not touch each other and that are usually ordered by height.
13. Joint bars may be used in bar graphs to compare data from two or more groups within each category.
14. The **line graph** is used to show trends and changes in values over time. The horizontal axis displays scores measured on a continuum (for example, years or months). The vertical axis can be used for various types of data (for example, test scores, temperatures, and income). A line connects the data points on the graphs.
15. A line graph can be used to compare multiple groups where each group on the graph is represented by its own line.
16. The **box plot** (also called **box-and-whiskers**) consists of a box and two whiskers and is used to depict the spread and skewness of frequency distri-butions. The box represents the middle 50 percent of the distribution, and

the whiskers represent the top and bottom 25 percent of the distribution. A horizontal line inside the box represents the median of the distribution and the lengths of the whiskers show the spread of the distribution.

17. To create the box plot, the scores are first ordered and divided into four quartiles, identified as Q_1, Q_2, Q_3, and Q_4. The two middle quartiles (Q_2 and Q_3) are located within the box, whereas the two extreme quartiles (Q_1 and Q_4) are displayed using vertical lines (the whiskers) outside the box.

18. Graphs should be clearly labeled and easy to read and interpret. It is important to choose the right scale for the graphs in order to provide an accurate visual representation of the data. Other guidelines for drawing graphs are also discussed in this chapter.

CHECK YOUR UNDERSTANDING

1. Create a *frequency distribution* with test scores from your class, or other data sources. (If real scores are unavailable, make up a set of scores.)

   a. Explain the nature of the scores and what they represent.

   b. Organize the scores as needed (from highest to lowest or from lowest to highest).

   c. Tally and group together scores that are the same and create *class intervals*, if needed.

   d. Choose and draw an appropriate graph to illustrate your data. (Hand-draw your graph or use a software program such as Excel.)

2. Choose another type of graph to illustrate another type of data, using the examples in the book (for example, pie graph or bar diagram). Explain your choice.

3. Find an example of graphic presentation of data in a newspaper, websites, or other types of electronic or printed report. Name the graph and explain why, in your opinion, it was chosen to illustrate the data.

4

Measures of Central Tendency

Chapter 4 discusses three measures of central tendency: *mode, median,* and *mean.* A measure of central tendency is a summary score that is used to represent a distribution of scores. Depending on the type of data you have collected, one or more of these measures will be appropriate. In this chapter, you will find definitions, explanations, and examples of each measure. Additionally, advantages and disadvantages of each measure are clearly presented and highlighted. The detailed guidelines that are included in this chapter about which measure to use in any given situation will help you choose the right measure.

In addition to describing distributions using frequency tables and graphics, we can also use a single representative score that will show the *center* of the distribution. There are three such measures of central tendency: the *mode*, median, and mean. In this chapter, we demonstrate how to compute them. However, you can easily obtain these statistics, along with other descriptive statistics, by using readily available computer programs (for example, Excel and SPSS).

MODE

The **mode** of a distribution is the score that occurs with the greatest frequency in that distribution. To illustrate the computation of the mode, let's examine the frequency column in table 4.1, which shows the scores of fifteen students. We can see that the score of 8 is repeated the most (four times); therefore, the mode of the distribution is 8.

Table 4.1.   Test Scores of Fifteen Students

	Score	Frequency
	12	1
	11	1
	10	2
	9	3
Mode →	8	4
	7	2
	6	1
	5	1

A score should repeat at least twice in order to be considered a mode. In a frequency polygon, the mode is the peak of the graph (figure 4.1, *part a*). If two scores have the same frequency, the distribution is called **bimodal**. Bimodal distributions have two peaks, both the same or similar height (figure 4.1, *part b*).

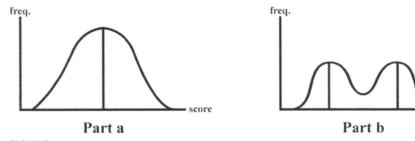

Part a   Part b

FIGURE 4.1
Frequency polygons with one mode (*part a*) and two modes (*part b*)

Table 4.2 shows an example of a *bimodal* distribution. The two modes are the scores of 5 and 4.

Table 4.2. A Bimodal Distribution with Two Modes (4 and 5)

	Score	Frequency
Mode →	6	1
	5	2
Mode →	4	2
	3	1
	2	1

If three or more scores repeat the same number of times, the distribution is referred to as a **multimodal** distribution. If no score is repeated, the distribution looks flat and is classified as **amodal** (has no mode).

MEDIAN

The **median** is the middle point of a distribution of scores that are ordered. Fifty percent of the scores are above the median, and 50 percent are below it. For example, in the distribution of scores in table 4.3, the score of 6 is the median because there are three scores above it and three below it.

Table 4.3. Test Scores of Seven Students with a Median of 6

	Score
	10
	8
	7
Median →	6
	4
	2
	1

The median is a *point*, and it does not have to be an actual score in that distribution. For example, suppose a special education teacher administers a weekly quiz to the four students in her class. The scores of the students are: 10, 8, 7, and 6. The median of the distribution is 7.5, even though the teacher assigns only full points on the test.

When the *exact* median is hard to calculate, it can be *estimated*. For example, the median of the seven scores in table 4.4 is estimated to be 8, even though the number of scores above is not exactly the same as the number of scores below it. As we can see, there are three scores above the median of 8 and two scores below it.[1]

---

1. HINT: Computer programs can provide the *exact* median of each distribution. There are also ways to compute the exact median by hand.

Table 4.4. Test Scores of Seven Students with a Median of 8

	Score
	13
	11
	9
Median →	8
	8
	7
	6

MEAN

The **mean**, which is also called the *arithmetic mean*, is obtained by adding up the scores and dividing that sum by the number of scores.[2] The mean, which is used in both descriptive and inferential statistics, is used more often than the mode or the median.

The statistical symbol for the mean is $\bar{x}$ (pronounced "ex bar"), and the symbol for the *population mean* is μ, the Greek letter *mu* (pronounced "moo" or "mew"). The statistical symbol for "sum of" is Σ (the capital Greek letter *sigma*). A **raw score** is a score as it was obtained on a test or any other measure, without converting it to any other scale. A *raw score* is represented in statistics by the letter X. ΣX means "the sum of all the X scores." Thus, the mean may be computed as ΣX divided by the number of scores:

$$\text{Mean} = \frac{\Sigma X}{N}$$

Where: ΣX = Sum of scores
            N = Number of scores

The mean serves as the best measure when we have to estimate an unknown value of any score in a distribution (for both samples and populations). That is, if the exact value of a particular score is unknown, the mean may be used to estimate that score.

In research, the mean of a sample ($\bar{x}$) is often used to estimate the population mean (μ). In many studies, researchers are interested in finding the mean of the population; however, it may not be practical or possible to study the whole population in order to find its mean. Therefore, they select a sample, measure it to obtain its mean, and use that mean to estimate the population mean.

COMPARING THE MODE, MEDIAN, AND MEAN

The mean is affected by *every* score in the distribution because to calculate the mean all the scores are first added before dividing that sum of scores (ΣX) by the number of scores (N). Changing even one score in the distribution may result in a change in the

---

2. HINT: The mean is sometimes called "the average," although the word *average* may also be used in everyday life to mean "typical" or "normal."

mean. By contrast, the mode and the median may, or may not, be changed as a result of a change in one score. This characteristic of the mean can be both an advantage and a disadvantage. It is an advantage because the mean is a measure that reflects every score in the distribution. It is a disadvantage when there are extreme scores in a skewed distribution. Let's look, for example, at two distributions A and B (table 4.5).

Table 4.5.   Comparing the Mode, Median, and Mean of Two Distributions

	Distribution A	Distribution B
	10	10
	12	12
	13	13
	13	13
	15	15
	16	40
Mode:	13	13
Median:	13	13
Mean:	13.7	17.17

Distribution A has six scores: 10, 12, 13, 13, 15, and 16. Its mode is 13, the median is 13, and the mean is 13.17. All three measures are similar to each other, and all can represent the distribution. Now, let's look at distribution B where we changed the last score from 16 to 40. This change has no impact on the mode or the median of distribution A when comparing to distribution B, but the mean of distribution B changes drastically from 13.17 to 17.17. The extreme score of 40 in the second distribution "pulled" the mean upward. Consequently, the mean of 17.17 does not represent any of the six scores in the second distribution. It is too high for the first five (10, 12, 13, 13, and 15) and is much too low for the last score of 40.

Not all three measures of central tendency can be used with all types of data. The mode is the only measure of central tendency that can be used with nominal scale data. As you recall, in nominal scales the observations are not ordered in any way (see chapter 2). Because the mode is an index of frequency, it can be used with observations that are not ordered. Mode can also be used with data measured on ordinal, interval, and ratio scales.

To find the median (the middle point), we need to be able to order the scores. A nominal scale has no order; therefore, the median cannot be used with nominal scale data. The median can be used with ordinal, interval, and ratio scales where scores can be ordered. The mean can be computed only for interval and ratio scale data because to calculate the mean, we need to add the scores and divide the sum by the number of scores.

The mode and the median are most often used for descriptive statistics, whereas the mean is used for descriptive statistics *and* inferential statistics. For example, the mean is used to compute the variance, standard deviation, and *z* scores (see chapters 5 and 6). It can also be used to compute other statistical tests such as the *t* test (see chapter 10). The median is used in everyday life for reporting incomes and housing prices.

When distributions of scores that are measured on an interval or ratio scale include extreme scores, the median is usually chosen as a measure of central tendency. For example, assume we have a few expensive homes in an area where most of the homes are moderately priced. The mean housing price may cause potential buyers, who rely on this information, to think that *all* the houses in that area are too expensive for them. Therefore, when the mean is inflated, it does not serve as a true representative score, and the median should be used.

SUMMARY

1. A **measure of central tendency** is a summary score that represents a set of scores. There are three commonly used measures of central tendency: *mode*, *median*, and *mean*.
2. The **mode** of a distribution is the score that occurs most frequently in that distribution.
3. A distribution of scores may have one mode, two modes (**bimodal**), three or more modes (**multimodal**), or no mode (**amodal**).
4. The **median** is the middle point of a distribution of scores that are ordered. Fifty percent of the scores are above the median and 50 percent are below it.
5. The **mean**, which is also called the *arithmetic mean*, is calculated by dividing ΣX (the total sum of the scores) by the number of scores (*N* or *n*). The symbol for the sample mean is $\bar{x}$ ("ex bar"); the symbol for the population mean is μ (the Greek letter *mu*).
6. The mean serves as the best measure when we have to estimate an unknown value of any score in a distribution (for both samples and populations).
7. In research, the mean of a *sample* $\bar{x}$ ("ex bar") is often used to estimate the *population* mean (μ).
8. The mean is affected by *every* score in the distribution because to calculate the mean all the scores are first added before dividing that sum of scores (ΣX) by the number of scores. Changing even one score in the distribution may result in a change in the mean. By contrast, the mode and the median may, or may not, be changed as a result of a change in one or a few scores.
9. The mode can be used with nominal, ordinal, interval, and ratio scales. The median can be used with ordinal, interval, and ratio scales. The mean can be used with interval and ratio scales.
10. The mean is not an appropriate measure of central tendency when interval or ratio scale distributions have extreme scores because it may yield a skewed measure. In such cases, the median, which is not affected by extreme scores, should be used.
11. The mode and the median are most often used for descriptive statistics, whereas the mean is used for descriptive statistics *and* inferential statistics. The mean can also be used to compute other statistical tests, such as the *t* test.

CHECK YOUR UNDERSTANDING

1. Why are there three measures of *central tendency*? What are the similarities and differences among them?

2. Give three examples of situations where you might use the *mode, median,* and *mean.*

3. Find an example of at least two measures of central tendency (mode, median, or mean) that are reported in the media (print or electronic); describe briefly how these are used in that context.

5

Measures of Variability

In addition to using a summary score to represent a distribution of scores, it is important to describe the variability, or spread, of scores in that distribution. Chapter 5 describes three measures of spread: *range*, *variance*, and *standard deviation*. Range is a fairly common measure that you are probably familiar with already. The variance is not an important measure on its own, but it is related to standard deviation and is used in the computations of a few other statistical tests. Standard deviation, by comparison, is an important concept to understand because it is used extensively in statistics and assessment. In essence, it describes the average deviation around the mean.

Although the computational steps of each measure are demonstrated in this chapter, you are likely to use computer programs to do the computations for you; therefore, make sure you understand the essence of each measure and do not worry about how to calculate them.

In chapter 4, we described a measure of central tendency (a mode, a median, or a mean) as a representative score; that is, a single number that represents a set of scores. These measures indicate the *center* of the distribution. In this chapter, we introduce you to several measures of *variability* that are useful in describing a distribution, especially when combined with measures of central tendency.

THE NEED FOR MEASURES OF VARIABILITY

To illustrate the need for measures of variability and spread, we present two graphs in figure 5.1. After examining this graph you will realize that, in addition to measures of central tendency, it is often necessary to obtain an index of the *variability* or *spread* of the group.

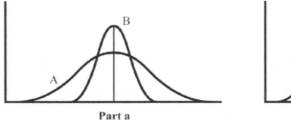
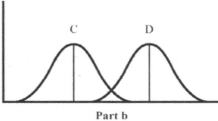

Part a   **Part b**

FIGURE 5.1

A graph showing two groups with the same means but different spreads (*part a*); and a graph showing two groups with different means but the same spreads (*part b*)

Note that the two groups in *part a* (groups A and B) share the *same mean*, but group A has a larger spread (that is, group A is more *heterogeneous* than group B). Group B seems to be more *homogeneous*, and this group's scores tend to cluster closer to the mean. Next, examine *part b* of figure 5.1. Notice that group C and group D have the same spread, or variability, but group D has a higher mean than group C.[1]

Suppose *part a* in figure 5.1 represents test scores of two third grade classes. If you were asked to choose which class you would like to teach and knew only that both classes have the same mean test scores, you might be tempted to flip a coin to choose the class you want to teach. If a graphic representation was available to you, as depicted in *part a*, you would probably choose the third grade class represented by B, which is more homogeneous and, therefore, easier to teach.

Looking at *part b*, assume the same question was asked and you were given only the means for both group C and group D. You would probably choose the third grade class depicted by D, which has a higher mean. In fact, even though the mean of group D is higher than the mean of group C, teaching either group would involve about the same amount of work for the teacher.

1. HINT: How can you tell which mean is higher if there are no numbers along the horizontal axis? Remember the rule about drawing a frequency polygon (discussed in chapter 3): the numbers increase as you move to the right on the horizontal axis. Since the mean of group D is farther to the right, it is higher than the mean of group C.

The examples in figure 5.1 were provided to demonstrate to you that the mean alone does not provide a complete and accurate description of a group. In addition to the mean, another index is needed in order to indicate the variability of the group. In this chapter, we discuss three measures of variability: *range, standard deviation,* and *variance.*

THE RANGE

The **range** indicates the distance between the highest and the lowest score in the distribution. The range is a simple and easy-to-compute measure of variability. To find the range, simply subtract the lowest score from the highest score in the distribution.

The range has limited usefulness as a measure of variability, and it does not give us much information about the variability *within* the distribution. The range is used much less frequently compared with the other two measures of variability discussed in this chapter (the variance and standard deviation).

To illustrate why range does not tell us much about the variability within a distribution of scores, compare these two sets of numbers: 10, 10, 10, 9, 1; and 10, 2, 1, 1, 1. Both distributions have the same range (10 – 1 = 9), yet the set of scores they represent is very different.

IQR = how spread out middle values are

STANDARD DEVIATION AND VARIANCE

The distance between each score in a distribution and the mean of that distribution $(X-\bar{X})$ is called the **deviation score**.[2] Looking at figure 5.2, we can expect higher deviation scores in *part a*, where the scores are widely spread, than *part b*, where most of the scores cluster around the mean. We would expect that in distributions with a high spread of scores, the mean (average) of the deviation scores would be higher than the distributions where most scores are closer to the mean.

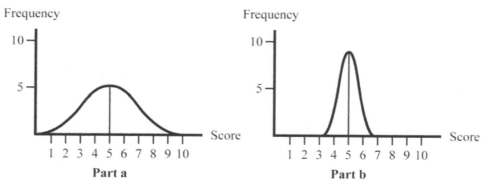

FIGURE 5.2
Two graphs showing distributions with the same means but different spreads

---

2. HINT: Some textbooks use the symbol x (lowercase "ex") to represent the deviation score. Other texts, including this textbook, use $\bar{x}$ to represent the deviation score.

The mean of the deviation scores is called the **standard deviation**, abbreviated often as SD. The SD describes the mean distance of the scores around the distribution mean. Squaring the SD gives us another index of variability, called the *variance*. As you will see next, the variance is needed in order to calculate the SD.

Let's look at the computational steps needed to calculate the deviation scores, variance, and SD. In order to simplify the computations, we will use a small distribution of five scores as an example. The numbers are: 6, 5, 4, 3, and 2 (see table 5.1).

To compute the deviation scores, we need to first compute the mean of the five scores. We do so by adding up the raw scores (ΣX) and dividing them by the number of scores ($n = 5$). The mean is 4.

$$\text{Mean} = \bar{X} = \frac{\sum X}{n} = \frac{20}{5} = 4$$

Table 5.1. Computing Deviation Scores

(Raw Scores)	(Deviation Scores)
X	$X - \bar{X}$
6	6 – 4 = 2
5	5 – 4 = 1
4	4 – 4 = 0
3	3 – 4 = -1
2	2 – 4 = -2
$\Sigma X = 20$	$\Sigma(X - \bar{X}) = 0$
Sum of Raw Scores	Sum of Deviation Scores

As you can see in table 5.1, the sum of the deviation scores is 0 (zero). Therefore, we cannot divide that sum by 5 (the number of scores) to find the mean of the deviation scores, as we were planning to do. As a matter of fact, with any combination of numbers, the mean of the deviation scores will always be 0 (zero). This is because the sum of the deviation scores *above* the mean of a distribution is equal to the sum of the deviation scores *below* the mean. We can overcome this problem by squaring each deviation score first, then finding the mean of these squared deviation scores.

The next step, then, is to square each deviation score, add up the squared deviation scores, and find their mean (see table 5.2). The mean of the squared deviations is the *variance*. Therefore, the **variance** may be defined as the mean of the squared deviations around the mean. The statistical symbol used to represent the sample variance is S^2.

Table 5.2. Computing the Variance

X	$X - \bar{X}$	$(X - \bar{X})^2$
6	6 – 4 = 2	4
5	5 – 4 = 1	1
4	4 – 4 = 0	0
3	3 – 4 = -1	1
2	2 – 4 =- 2	4
$\Sigma X = 20$	$\Sigma (X - \bar{X}) = 0$	$\Sigma (X - \bar{X})^2 = 10$

It is important to know whether our distribution of scores is a *sample* or a *population* because there is a difference in the computational steps for finding the variances of samples and populations. The difference is in the denominator of the formula used to compute the variance. The denominator is $n - 1$ (number of scores minus 1) for samples, and N (number of scores) for populations. Because there are five scores in our example, we assume the scores to be a *sample* rather than a *population*.

The formula for the computation of the sample variance (S^2) is:

$$\text{Variance} = S^2 = \frac{\sum (X - \bar{X})^2}{n-1}$$

Where S^2 = Sample variance

$\Sigma (X-\bar{X})$ = Sum of the squared deviations from the mean

$n - 1$ = Number of scores minus 1

Replacing the symbols with our numbers from table 5.2 will give us a sample variance of 2.5.

$$\text{Variance} = S^2 = \frac{\sum (X - \bar{X})^2}{n-1} = \frac{10}{4} = 2.5$$

Our reason for computing a measure of variability was that it would provide us an index of the mean (average) of the deviations around the mean. Clearly, the variance is not a good measure for that purpose. In our example, we computed a variance of 2.5, which is not representative of the deviation scores of 2, 1, 0, –1, and –2. The reason why the variance is higher than all the deviation scores in our example is pretty obvious: As you recall, the sum of the deviation scores in table 5.1 was 0. To overcome this problem (see table 5.2), we *squared* the deviation scores. Therefore, to get over the problem of obtaining such an inflated index, we simply reverse the process. In other words, we find the *square root* of the variance, which would help us get back to the original units we used in table 5.1. The square root of the variance is the *standard deviation* (SD, or *S*). In our example, the variance is 2.5 and its square root, the SD, is 1.58. The computation formula for the standard deviation is:

$$S = \sqrt{VAR} = \sqrt{S^2} = \sqrt{2.5} = 1.58$$

Where S = Sample standard deviation
S^2 = Sample variance

Conversely, we can reverse the process and find the variance by *squaring* the SD. For example, when the SD is 5, we can find the variance (S^2) by squaring the SD.

$$S^2 = SD^2 = 5^2 = 25$$

Where S^2 = Sample variance
SD^2 = Sample SD, squared

Because the SD is the square root of the variance, it is usually *smaller* than the variance. However, this is not always the case. As you know, the square root of 1 is 1. In fact, when the variance is less than 1, the SD will be *higher* than the variance. For example, when the variance is 0.8, the SD is 0.89 and when the variance is 0.5, the SD is 0.7.

While the symbols that are used for the sample standard deviation and variance are S and S^2, respectively, Greek letters are used for the population values. The symbol used for the population standard deviation is σ (Greek lowercase letter sigma), and the symbol for the population variance is σ^2.

Computing the Variance and SD for Populations and Samples

As was mentioned earlier, there is a difference in the formulas used to compute the sample and population variances and standard deviations. Following is an explanation of that difference.

The standard deviation of a population (σ) is a fixed number, but the sample standard deviation (*S*) varies, depending on the sample that was selected. If we select several samples from the same population and compare their standard deviations, we are likely to see that not all of them are exactly the same.[3] When researchers started comparing such samples to the population from which they were selected, they realized that not all samples have the same means and standard deviations. Further, researchers also found that variances and standard deviations from samples were consistently *lower* than the variances and standard deviations of the populations from which the samples were selected. This was especially true with small samples (with $n \leq$ 30). In conducting research, researchers often use the variance and standard deviation from a single sample to *estimate* the variance and standard deviation of the popula-

---

3. HINT: The same applies to means: Means from different samples selected from the same population are likely to differ, whereas the mean of the population is a fixed number (see chapter 2).

tion. Because the sample variance and standard deviation are likely to consistently *under*estimate the population variance and standard deviation, there was a need to modify the equations used to compute the sample variance and standard deviation. These modified equations result in a slightly higher variance and standard deviation that are more representative of the variance and standard deviation of the population.

For example, in table 5.2 we have only five scores. It is very likely that such a small group of scores is a sample, rather than a population. Therefore, we computed the variance and SD for these scores, treating them as a *sample*, and used a denominator of $n - 1$ in the computations that followed table 5.2. When we consider a set of scores to be a population, we should use a denominator of N to compute the variance.

The difference in the choice of denominator is especially important when the number of cases in the distribution is small. Consider, for example, the data in table 5.2. In our computation of the variance, which follows that table, we had a numerator of 10 and used a denominator of 4 (n-1). If we had decided that the group of five scores is a population, we should have used a denominator of N when computing the variance. Consequently, with a numerator of 10, we would have obtained a variance of 2 instead of our variance of 2.5 and a SD of 1.41 instead of our SD of 1.58.

When the sample is large, the choice of the proper denominator (N versus n-1) to be used for the computation of the variance and SD is not as important as when the sample is very small. For example, let's assume we have a set of 100 scores and the numerator in the formula used to compute the variance is 800. If that set of scores is considered a *population*, then to compute the variance we will divide the numerator by 100. The variance would be 8.00 (800 divided by 100). The SD for that population of 100 scores would be 2.83 (the square root of the variance). If the set of scores is considered to be a *sample*, then the denominator would be 99 (n-1) and the variance would be 8.08 (800 divided by 99). The SD would be 2.84 (the square root of 8.08). Clearly, there is very little difference between the two standard deviations (2.83 for a population and 2.84 for a sample) when the sample size is large.

In real life, you are not likely to have to calculate either the variance or the SD by hand. There are many computer programs that are easy to use that can calculate both of these for you. We introduced the computation steps here simply to explain these concepts.

Using the Variance and SD

A higher variance may show higher variability in a group, compared with a lower variance, but these values are difficult to interpret because they are measured in *squared* units. The SD is measured in the same units as the original data and is easier to interpret. For example, when measuring height, a SD of 3 means that, on average, the heights of the members of the group deviate 3 inches from the mean.

The variance may be used as an intermediate step in the computation of the standard deviation. The variance is also found in the computational steps of some statistical tests such as *t* test (see chapter 9) and analysis of variance (ANOVA) (see chapter 10). Standard deviation is often used to summarize data, along with the mean or other measures of central tendency. For instance, in reporting results of tests, we are most likely to use summary scores, such as the mean and standard deviation. Technical manuals for tests show extensive use of the mean and SD. Further, scales of tests are usually described in terms of their mean and SD. For example, we are told that a certain IQ test has a mean of 100 and a SD of 15. (These concepts are presented in chapter 6, which discusses the normal curve, and chapter 12, which discusses standardized test scores.)

Variance and SD in Distributions with Extreme Scores

The variance and the SD are sensitive to extreme scores. Having skewed distributions with even one extreme score may substantially increase the variance and SD. Consider these two sets of scores, set 1 and set 2 (table 5.3). Note that the two sets are the same with the exception of one extreme score in set 2 (40, the first score in set 2).

Table 5.3. Two Sets of Scores, with an Extreme Score in Set 2

	Set 1	Set 2
	11	40
	10	10
	9	9
	8	8
	7	7
	6	6
	5	5
	4	4
	3	3
	2	2
$\Sigma X =$	65.00	94.00
$\bar{X} =$	6.50	9.40
VAR =	9.17	122.27
SD =	3.03	11.06

Set 1 has a SD of 3.03, which seems like a good representation of the distances of the scores around their mean of 6.50. The mean of 6.50 in set 1 is also a good representation of the scores in that set. By comparison, the SD of 11.06 in set 2 is much higher than the SD in set 1, due to the extreme score of 40. The mean of 9.40 in set 2 is also misleading, and it is higher than most of the scores in that set.

The SD is supposed to be an index of the average distances of the scores around their mean. In set 2, the SD of 11.06 provides misleading information. It implies a much higher variability in the group, compared with the SD in set 1. In fact, all the scores in set 2, with the exception of the first score of 40, are fairly close together.

Factors Affecting the Variance and SD

As you probably have noticed, there is a relationship between the range, variance, and SD; the wider the range, the higher the variance and SD. The range is higher when the group is more *heterogeneous* regarding the characteristic being measured. This characteristic can be, for example, height, IQ, reading scores, and age. The range, variance, and SD are also higher when there is at least one extreme score in the distribution, even if the rest of the scores cluster together (see set 2 in table 5.3). These three measures (range, variance, and SD) tend to be lower when the scores cluster together, as is the case in set 1 in table 5.3.

The length of a test can also affect the variance and SD. A longer test has the potential to spread the scores more widely and thus have a higher SD than does a short test. Compare, for example, two tests: test A, with 100 items, and test B, with 10 items. Let's assume the mean of test A is 50, and the mean of test B is 5. In test A, people might score up to 50 points above or below the mean. In test B people might score only up to 5 points above or below the mean. Since the SD is a measure of the mean (average) distance of the scores from the mean, it is likely to be higher in test A than in test B. The same would apply to the variance of the distribution of scores, which is obtained by squaring the SD of that distribution. Shorter tests tend to produce smaller variances than longer tests.

Another factor that can affect the variance and SD is the level of difficulty of a test. When a test is very easy, most students answer all the questions correctly; therefore, the scores cluster together, there is little variability, and the variance and SD are likely to be lower. Clustering of the scores can happen in mastery tests and criterion-referenced tests that tend to be easier. Similarly, scores from tests that are very difficult for all examinees tend to cluster together at the low end of the distribution. When scores cluster at the high end or the low end of the distribution curve, the variance and SD tend to be lower than in cases where the scores are spread along a bell-shaped distribution. Norm-referenced commercial tests are designed to spread the scores widely to create a bell-shaped distribution (see chapter 12). Scores from such tests would have higher variance and SD than those obtained on tests where the scores tend to cluster together.

SUMMARY

1. To describe a distribution of scores, an index of *variability*, as well as a measure of central tendency, is needed.

2. The **range** is the distance between the highest and the lowest scores in the distribution. To calculate the range, subtract the lowest score from the highest score.
3. The range is an index of the variability of the group, and it is used mostly for descriptive purposes.
4. The **deviation score** is the distance of the raw score from the mean, indicated by $x - \bar{x}$ (that is, the score minus the mean).
5. The sum of the deviation scores (that is, the distances between the raw scores and the mean of that distribution) is always 0 (zero).
6. The **variance** is the mean of the squared deviations. To calculate it, square each deviation score, add all the squared deviations, and divide their sum by $n - 1$ (the number of scores minus 1) for the *sample* variance.
7. In the equation used to compute the *population* variance, the denominator is N (number of scores). Choosing the correct equation is especially important when the sample size is small ($n \leq 30$).
8. The **standard deviation** (**SD**) is the mean (average) distance of scores from the mean. It can be computed by finding the square root of the variance:

$$SD = \sqrt{Variance}$$

9. When we square the SD, we can find the variance:

$$Variance = SD^2$$

10. The symbol that is used for the *sample* variance is S^2, and the symbol used for the *population* variance is σ^2. The symbol for the *sample* standard deviation is S, and the symbol for the *population* standard deviation is σ (the Greek lowercase letter *sigma*).
11. The standard deviation of a population (σ) is a fixed number, but the sample standard deviation (S) varies, depending on the sample that was selected.
12. Empirical research has shown that the variances and standard deviations of small samples consistently *under*estimate the population variance and standard deviation. Therefore, the equations used to compute the variance and standard deviation of samples were modified to produce slightly higher variances and standard deviations that are more representative of the variance and standard deviation of the population.
13. The standard deviation is measured using the same units as the original data and is easier to interpret than the variance. Standard deviation is often used along with the mean in summarizing and reporting test data.
14. The variance is not commonly used when describing a distribution of scores. The reason is that the variance, which is expressed in *squared* units, tends to be much larger than the majority of the deviation scores around the mean of the distribution.
15. The range, variance, and standard deviation are sensitive to extreme scores. Using the same test, groups with a wide range of scores (heterogeneous

groups) have a larger range, variance, and standard deviation than groups where scores cluster together (homogeneous groups).

16. Variance and standard deviation of longer tests (with more items) tend to be higher than the variance and standard deviation of shorter tests (with fewer items).

17. In tests that are very easy or very difficult, the scores of examinees tend to cluster at one end and the variance and standard deviation are likely to be low on such tests.

18. In norm-referenced commercial tests, the examinees' scores are usually spread along a bell-shaped curve. Therefore, the scores tend to have a wider range, resulting in higher variance and standard deviation.

CHECK YOUR UNDERSTANDING

1. What is the difference between *variance* and *standard deviation*? Explain.
2. How do the variability and range of a distribution of scores affect the distribution's standard deviation? Explain.
3. Which measure is more useful in reporting on test scores: the variance or the standard deviation? Explain.
4. Suggest at least one example from your educational setting or other settings where *standard deviation* and *means* are reported or may be reported. Explain your choices.

6

The Normal Curve and Standard Scores

Chapter 6 introduces the *normal curve, skewed distributions,* and *standard scores*. The normal curve is important as a theoretical and practical model, and its unique features are discussed and explained. You are probably familiar with the bell-shaped distribution, which is also a normal curve, and know that various characteristics in nature (for example, height and IQ) are distributed in a bell-shaped distribution. We also show you distributions that are asymmetric where the majority of the scores cluster either on the right or left side.

Using the bell-shaped distribution, you can convert raw scores into standardized scores, such as *z scores* and *percentile ranks*. Doing so allows you to compare scores obtained on different tests, each with its own mean and standard deviation. For example, we can compare how well a student has performed on reading and math tests if we know the student's scores on these two tests, and the means and standard deviations of the whole class on the two tests. Advantages and appropriate uses for each type of standard score are also included in this chapter.

This chapter introduces you to the concept of the normal curve as a theoretical model and to several applications of this model to the field of education. We describe the model and its unique features, and show you a variety of normal distributions, as well as skewed distributions. Examples of standard scores are also discussed in the chapter.

THE NORMAL CURVE

For years, scientists have noted that many variables in the behavioral and physical sciences are distributed in a bell shape. These variables are *normally distributed* in the population, and their graphic representation is referred to as the **normal curve**.[1] For example, in the general population, the mean of the most commonly used measure of IQ is 100. If the IQ scores of a group of 10,000 randomly selected adults are graphed using a frequency polygon, the graph is going to be bell-shaped, with the majority of people clustering just above or below the mean. There would be increasingly fewer and fewer IQ scores toward the right and left tails of this distribution as the IQ scores get higher or lower. Similarly, if we were to record the heights of 1,000 three-year-old children and then graph the distribution, we would see that it forms a normal curve.

The development of the mathematical equation for the normal distribution is credited, according to some sources, to the French mathematician Abraham de Moivre (1667–1754). According to other sources, it was the German mathematician Karl Friedrich Gauss (1777–1855) who developed the equation. Thus, the normal curve is also called the "Gaussian Model."

The normal curve is a theoretical, mathematical model that can be represented by a mathematical formula. Because many behavioral measures are distributed in a shape like the normal curve, the model has practical implications in the behavioral sciences and education. In this chapter, we show how this model can be applied to education.

The **normal distribution** is actually a group of distributions, each determined by a mean and a standard deviation. Some of these distributions are wider and more "flat," while others are narrower, with more of a "peak" (see figure 6.1).

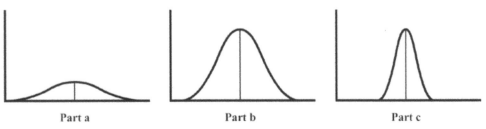

Part a   Part b   Part c

FIGURE 6.1
Three normal distributions with different levels of "peakedness" (or "flatness")

---

1. HINT: The standard normal curve is referred to by most people as the normal curve, or a bell-shaped distribution.

Regardless of the exact shapes of the normal distributions, all share four characteristics:

1. The curve is symmetrical around the vertical axis (half the scores are on the right side of the axis, and half the scores are on the left).
2. The scores tend to cluster around the center (that is, around the mean, or the vertical axis).
3. The mode, median, and mean have the same values.
4. The curve has no boundaries on either side (the tails of the distribution are getting very close to the horizontal axis, but never quite touch it).[2]

Although many characteristics are normally distributed in the population, measuring and graphing these characteristics for a small number of cases will not necessarily look like the normal curve. *Part a* (in figure 6.2) depicts a distribution of scores obtained from a smaller sample, and *part b* depicts scores from a larger sample. Note that the graph of the distribution in *part b* looks "smoother" than the graph of the distribution in *part a*. The reason is that as the number of cases increases, the shape of the distribution is more likely to approximate a normal curve.

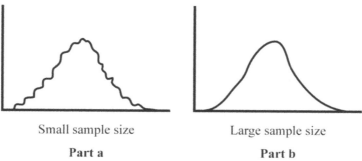

Small sample size   Large sample size

Part a   **Part b**

FIGURE 6.2
Normal curve distributions with a smaller sample size (*part a*) and a larger sample size (*part b*)

The normal curve is divided into segments, and each segment contains a certain percentage of the area under the curve (see figure 6.3). The distances between the various points on the horizontal axis are equal, but the segments closer to the center contain more scores than the segments farther away from the center.

Figure 6.3 shows the normal curve with a distribution of scores that have a mean of 0 and a standard deviation (SD) of 1. The units *below* the mean (on the left side) are considered *negative* (for example, –1, –2), and the units *above* the mean (on the right side) are considered *positive* (for example, +1, +2).

---

2. HINT: Keep in mind that this is a theoretical model. In reality, the number of scores in a given distribution is finite, and certain scores are the highest and the lowest points of that distribution.

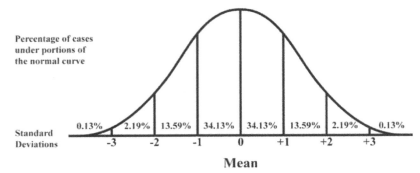

FIGURE 6.3
A bell-shaped (normal) distribution with a mean of 0 and a standard deviation of 1

In normal distributions, 34.13 percent of the scores are expected to be between the mean and +1SD and 34.13 percent of the scores are expected to be between the mean and –SD. The area between the mean and 2SD above the mean or between the mean and –2SD is expected to include 47.72 percent (34.13 + 13.59 = 47.72) of the scores (see figure 6.4). The area between 3SD above and 3SD below the mean is expected to contain almost all the cases in the distribution, 99.74 percent.

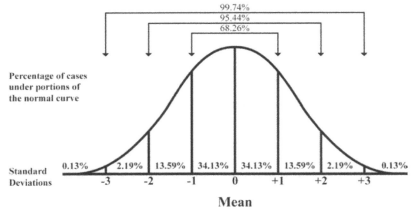

FIGURE 6.4
The percentages of the area under the normal curve at ±1SD, ±2SD, and ±3SD

The normal curve can be used to describe, predict, and estimate many types of variables that are normally distributed. If we know the distribution mean and standard deviation, we can estimate the percentages of scores in different parts of the distribution. For example, we can use the normal curve to estimate the IQ scores of people in the general population by using the information about the areas under the normal curve.

Suppose we use an IQ test with a mean of 100 and a standard deviation of 15. Using the information in figure 6.5, we can mark the mean IQ score as 100, the IQ score of 115 at +1SD, and the IQ score of 130 at +2SD. In the area below the mean, –1SD corresponds to an IQ score of 85 and –2SDs correspond to an IQ score of 70. We can determine that in the general population, approximately 34 percent of the people are expected to have IQ scores between 100 and 115 (between the mean and +1SD), and about 68 percent (or two-thirds) of the people in the population are expected to have IQ scores between 85 and 115. (An IQ of ±1SD is considered within the normal range.)

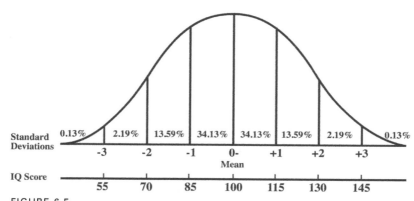

FIGURE 6.5
A bell-shaped distribution of IQ scores with a mean of 100 and SD of 15

In some school districts, students have to score at least +2SD above the mean on an IQ test to be defined as *gifted*. Using this criterion, we can ascertain that about 2 percent of the students in the general population would have IQ scores high enough to be classified as gifted.

SKEWED DISTRIBUTIONS

While in *normal* distributions there are approximately the same number of scores on the right and left side of the distribution midpoint, there are other *skewed* distributions with a longer tail on one side. A **skewed distribution** refers to an asymmetrical distribution where the majority of scores cluster either below or above the mean. In *negatively skewed* distributions, the long tail of the distribution is on the left (negative) side of the distribution where the lower scores are, whereas in a *positively skewed* distribution, the long tail of the distribution points toward the right-hand side, in the direction of the high scores.

To illustrate this point, let's look at figure 6.6 where the vertical lines in each part show the mean of the distribution. *Part a* shows a *symmetrical* bell-shaped distribution,

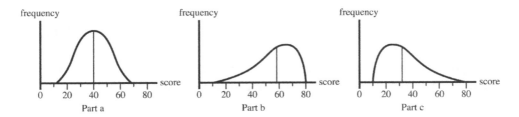

FIGURE 6.6

Symmetrical distribution (*part a*), negatively skewed distribution (*part b*), and positively skewed distribution (*part c*)

Part b in figure 6.6 depicts a *negatively skewed* distribution, and *part c* depicts a *positively skewed* distribution. As we can see, in these skewed distributions, the mean is pulled toward the "tail" and does not represent a point around which scores tend to cluster. Note that in a negatively skewed distribution the scores tend to cluster *above* the mean, and in a positively skewed distribution the scores tend to cluster *below* the mean.

Skewness may also be described in terms of the location of the long tail of the distribution. Thus, negatively skewed distributions may also call *skewed to the left*, because the tail is on the left side pointing toward the lower scores. Similarly, positively skewed distributions may be referred to as *skewed to the right*.

STANDARD SCORES

Until now, two types of scores were introduced in the book: *individual scores* (raw scores) and *group scores* (mode, median, mean, range, variance, and standard deviation). Raw scores are scores obtained by individuals on a certain measure, and group scores are summary scores that are obtained for a group of scores. Both types of scores are scale specific and cannot be used to compare scores on two different measures, each with its own mean and standard deviation. To illustrate this point about the limitation of raw scores and group summary scores, let's look at the following example.

Suppose we want to compare the scores obtained by a student on two achievement tests, one in English and one in mathematics. Let's say that the student received a score of 50 in English and 68 in mathematics. Because the two tests are different, we cannot conclude that the student performed better in mathematics than in English. Knowing the student's score on each test will not allow you to determine on which test the student performed better. We do not know, for example, how many items were on each test, how difficult the tests were, and how well the other students did on the tests. Simply put, the two tests are not comparable.

To be able to compare scores from different tests, we can first convert them into standard scores. A **standard score** is a derived scale score that expresses the distance of the original score from the mean in standard deviation units. Once the scores from different tests are converted into a common scale, they can then be compared to each other. Two types of standard scores are discussed in this chapter: *z scores* and *T scores*.[3]

z Scores

The **z score** is a type of standard score that indicates how many standard deviation units a given score is *above* or *below* the mean for that group. The z scores create a scale with a mean of 0 and a standard deviation of 1. The shape of the z score distribution is the same as that of the raw scores used to calculate the z scores.

The theoretical range of the z scores is $\pm \infty$ ("plus/minus infinity"). Since the area above a z score of +3 or below a z score of –3 includes only 0.13 percent of the cases, for practical purposes most people only use the scale of –3 to +3. To convert a raw score to a z score, the raw score as well as the group mean and standard deviation are used. The conversion formula is:

$$Z = \frac{Raw\ Score - Mean}{SD} \qquad Or \qquad Z = \frac{X - \bar{X}}{S}$$

Where X = Raw score

$\bar{X}$ = Group mean

S = Group standard deviation (SD)

Table 6.1 presents the raw scores of one student on four tests (social studies, language arts, mathematics, and reading). The table also displays the means and standard deviations of the student's classmates on these tests and shows the process for converting raw scores into z scores.[4]

Table 6.1. Student's Scores, Class Means, Class Standard Deviations, and z Scores on Four Tests

Subject	Raw Score	Mean	SD	Z Score
Social Studies	85	70	14	$\frac{85 - 70}{14} = +1.07$
Language Arts	57	63	12	$\frac{57 - 63}{12} = -0.50$
Mathematics	55	72	16	$\frac{65 - 72}{16} = -0.44$
Reading	80	50	15	$\frac{80 - 50}{15} = +2.00$

3. HINT: The *T* scores are not related to the *t* test that is discussed in chapter 9.
4. HINT: Computer programs, such as SPSS, can easily convert raw scores into z scores.

The raw scores that are *above* the mean convert into *positive z* scores, and the raw scores that are *below* the mean convert into *negative z* scores. Consequently, about half of the students are expected to get positive *z* scores and half are expected to get negative *z* scores. As we can see in table 6.1, a student may answer many questions correctly (for example, see the score of 85 on the social studies test), yet get a *z* score of 1.07, which may appear to be a low score. It is clear that for reporting purposes, *z* scores are not very appealing.

T Scores

The **T score** is another standard score measured on a scale with a mean of 50 and a SD of 10 (figure 6.7). In order to calculate *T* scores, *z* scores have to be calculated first. Using this standard score overcomes problems associated with *z* scores. All the scores on the *T* score scale are positive and range from 10 to 90. Additionally, they can be reported in whole numbers instead of decimal points.

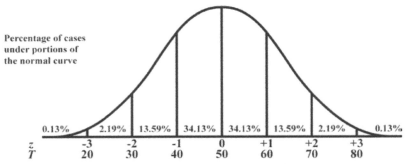

FIGURE 6.7
The normal curve showing z scores and corresponding T scores

In order to convert scores from *z* to *T*, we multiply each *z* score by 10 and add a constant of 50 to that product. The formula to convert from a *z* score to a *T* score is:

$$T = 10(Z) + 50$$

Next, we convert the *z* scores in table 6.1 into *T* scores using the conversion formula. The computations are displayed in table 6.2. *T* scores are usually rounded off and reported as whole numbers. Inspecting the table, we can see that negative *z* scores convert to *T* scores that are below 50. For example, a *z* score of –0.50 is converted to a *T* score of 45. Positive *z* scores convert to *T* scores that are higher than 50. For example, a *z* score of +2.00 is converted to a *T* score of 70.

Table 6.2.   Conversion of z Scores to T Scores

Subject	z Score	T Score
Social Studies	+1.07	10 (+1.07) + 50 = 60.7 or 61
Language Arts	−0.50	10 (−0.50) + 50 = 45.0 or 45
Mathematics	−1.06	10 (−1.06) + 50 = 39.4 or 39
Reading	+2.00	10 (+2.00) + 50 = 70.0 or 70

Other Converted Scores

Many measures used for educational and psychological testing indicate the position of an individual in relation to the *population*. The population is described in terms of mean and standard deviation. For example, the Wechsler IQ test has a mean of 100 and a SD of 15, and the Stanford-Binet IQ test has a mean of 100 and a SD of 16. The mean and SD of the ACT tend to fluctuate slightly but are likely to be 21.1 and 5.0, respectively.

The Normal Curve and Percentile Ranks

A **percentile rank** of a score is defined usually as the percentage of examinees that scored *at* or *below* that score. For example, a percentile rank of 65 (P_{65}) means that 65 percent of the examinees scored at or below that score. Another definition of a percentile rank is that it indicates the percentage of examinees that scored *below* that score (omitting the word "at"). The second definition is the one used most often by commercial testing companies on their score reports.

In practice, a percentile rank of 100 is not reported. We cannot say that a person with a certain raw score did better than 100 percent of the people in the group, because that person has to be included in the group. Instead, 99 percent (or in some cases, 99.9 percent) is considered the highest percentile rank.

Percentiles are used to describe various points in a distribution. For example, a percentile rank of 70 (P_{70}) is said to be at the 70th percentile. Since percentiles represent an ordinal, rather than interval or ratio scale, they should not be manipulated (for example, added or multiplied). If manipulation is desired, percentiles should first be converted to z scores (which have equal intervals) or to raw scores.

The normal curve can be used to calculate percentiles, assuming that the distribution of scores is normally distributed (see figure 6.8). For example, a z score of +1 corresponds to a percentile rank of 84.14 (or 84). We find that percentile rank by adding up the percent of scores between the mean and a z score of +1 on the normal curve (it is 34.14 percent) to the percent of scores below the mean (50 percent). A z score of −2 corresponds to a percentile rank of 2 (the percent of area under the normal curve below a z score of −2).

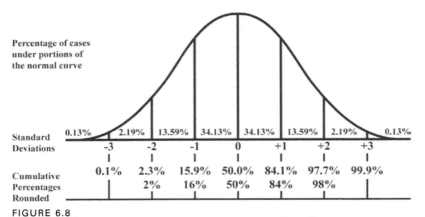

FIGURE 6.8
The normal curve with standard deviations and cumulative percentages

In most instances where we have a small group size ($n < 30$), the shape of the distribution will not approximate the normal curve. When the shape of the distribution is not assumed to be normal, it is inappropriate to use the normal curve model as a means to calculate percentiles. Instead, the teacher can use a simple formula to convert raw scores to percentile ranks. The information needed to convert a raw score to percentile rank is the number of students who scored below that score, the number of students who obtained the same score, and the total number of students. The formula is:

$$PR = \frac{N_{Below} + N_{at}}{N_{Total}}(100)$$

Where PR = Percentile Rank
       N_{Below} = Number of students who scored below that raw score
       N_{at} = Number of students who received the same score
       N_{Total} = Total number of students

For example, assume that three students in a class of twenty-five get a score of 21, scoring better than eighteen other students in the class. Using this formula, we obtain a percentile rank of 84 for these students. In other words, they did better than, or as well as, 84 percent of their classmates who took the same test. The computations are:

$$PR_{25} = \frac{18+3}{25}(100) = (0.84)(100)$$

In practice, you are not likely to compute the percentile ranks by hand. Instead, you can use computer programs (such as SPSS) to compute percentile ranks as well as z scores.

SUMMARY

1. The **normal curve** is a graphic representation of *normally distributed* variables in the behavioral and physical sciences.
2. The visual representation of the normal curve is bell shaped, with the majority of scores clustering just above or below the mean and increasingly fewer scores at either end of the curve.
3. The normal curve is a theoretical, mathematical model that can be represented by a mathematical formula. Since many behavioral measures are normally distributed, the model has practical implications in the behavioral sciences and education.
4. The **normal distribution** consists of a group of distributions, each determined by a mean and a standard deviation. Some of these distributions are wider and more "flat," while others are narrower, with more of a "peak."
5. The normal distribution has four characteristics: (a) it is symmetrical around the vertical axis; (b) the scores tend to cluster around the center; (c) the mode, median, and mean have the same value; and (d) theoretically, the curve has no boundaries on either side.
6. The normal curve is divided into segments, and each segment contains a certain percentage of the area under the curve. The distances between the various points on the horizontal axis are equal, but the segments closer to the center contain more scores than the segments farther away from the center.
7. In a normal distribution, 34.13 percent of the scores are expected to be between the mean and 1SD, and between the mean and –1SD. The area between the mean and 2SD above the mean is expected to include 47.72 percent of the scores, and the area between 3SD above and 3SD below the mean is expected to contain almost all the cases in the distribution (99.74 percent).
8. The normal curve can be used to describe, predict, and estimate many types of variables that are normally distributed. If we know the distribution mean and standard deviation, we can estimate the percentages of scores in different parts of the distribution.
9. A **skewed distribution** refers to an asymmetrical distribution where the majority of scores cluster either below or above the mean. In *negatively skewed* distributions, the long tail of the distribution is on the left (negative) side of the distribution where the lower scores are; whereas in a *positively skewed* distribution, the long tail of the distribution points toward the right-hand side, in the direction of the high scores.
10. A **standard score** is a derived scale score that expresses the distance of the original score from the mean in standard deviation units. Standard scores, such as *z* scores, can be used to compare raw scores from different distributions of scores (for example, from different achievement tests).
11. A **z score** is a commonly used standard score that indicates how many standard deviation units a given score is above or below the mean for that group. The group's mean and standard deviation are used to convert the raw scores to *z* scores. The conversion formula is:

$$Z = \frac{Raw\ Score - Mean}{SD}$$

12. Raw scores that are *above* the mean convert into *positive z* scores, and raw scores that are *below* the mean convert into *negative z* scores. Therefore, if classroom teachers convert their students' raw scores to *z* scores, the raw scores of approximately half of the students in the class are expected to convert to positive *z* scores and the other half to negative *z* scores. Students who score *exactly* at the mean, the most "average" students, are assigned a *z* score of 0.00.

13. Using *z* scores for the purpose of reporting students' scores can be problematic because students may be assigned negative scores, scores of 0, and scores with decimal places. Additionally, no student may get a score higher than 4.

14. **T scores** are standard scores that can range from 10 to 90, with a mean of 50 and a standard deviation of 10. To obtain *T* scores, it is necessary to find the *z* scores first. The conversion formula is: $T = 10(z) + 50$.

15. A **percentile rank** of a score is defined by most people as the percentage of examinees that scored *at* or *below* that score. For example, a percentile rank of 65 (P_{65}) means that 65 percent of the examinees scored at or below that score. Other definitions of a percentile rank state that it indicated the percentage of examinees that scored *below* a given score (omitting the word "at").

16. *Percentiles* are used to describe various points in a distribution. For example, a percentile rank of 70 (P_{70}) is said to be at the 70th percentile. Since percentiles represent ordinal, rather than interval or ratio scale, they should not be manipulated (for example, added or multiplied).

17. When the shape of the distribution is not assumed to be normal, percentile ranks can be calculated using this formula:

$$PR = \frac{N_{Below} + N_{at}}{N_{Total}}(100)$$

CHECK YOUR UNDERSTANDING

1. List two characteristics that are *normally distributed* in the general population.
2. List and explain at least two differences between *normal* and *skewed* distributions.
3. List and explain at least two differences between *z scores* and *T scores*.
4. Name at least one example where educators may use *raw* scores; explain why it would be appropriate to use raw scores in this example.
5. Name at least one type of *standard scores* that are used in education. Explain how they are computed and how they are reported.

III

MEASURING
RELATIONSHIPS

11

MEASURING
RELATIONSHIPS

Correlation

Chapter 7 is the first chapter that introduces you to specific statistical tests. In this chapter, we highlight *correlation*, a statistical test that is designed to study relationship and association between variables. Examples may include the relationship between reading and writing, school attendance and grade point average, and attitudes and motivation. A *correlation coefficient* is used to quantify and represent the relationship, and it can tell us whether the variables have a positive or negative correlation and whether the correlation is low, moderate, or high.

In this chapter, we use the *Pearson correlation coefficient,* the most commonly used correlation. After obtaining the coefficient, the next step is to determine whether the results are statistically significant or could have happened purely by chance. The concept of *level of significance*, or *p value*, is introduced in this chapter. It also contains an explanation of the process and a discussion of how to evaluate the *statistical* and *practical* significance of the computed correlation.

The chapter provides explanations on how to interpret the correlation coefficient and how to create a scattergram to display the correlation graphically. Although computers can easily do the computations for you, you need to decide when to use the correlation and how to interpret the results.

When there are more than two variables that are correlated, the results of the correlation are often displayed in correlation tables. Therefore, this chapter also includes an explanation of how to construct and read such tables.

The word *correlation* is used in everyday life to indicate a relationship or association between events or variables. However, in statistics, correlation refers specifically to the procedure used to *quantify* the relationship between two *numerical* variables through the use of a correlation coefficient. The following is an explanation of the concept of correlation and its uses in general, and specifically in education.

DEFINING CORRELATION

Correlation is defined as the relationship or association between two or more numerical variables. These variables have to be related to each other or paired. The most common way to use correlation in the field of education is to administer two measures to the same group of people and then correlate their scores on one measure with their scores on the other measure.

The strength, or degree of correlation, as well as the direction of the correlation (positive or negative), is indicated by a **correlation coefficient**. The coefficient can range from –1.00, indicating a perfect negative correlation; to 0.00, indicating no correlation; to +1.00, indicating a perfect positive correlation.

It is important to understand that correlation does not imply *causation*. Just because two variables correlate with each other does not mean that one caused the other. The only conclusion we can draw from a correlation between two variables is that they are related. In many cases, there is a possibility that there is a third variable that causes both variables to correlate with each other.

In addition to being used to describe the relationship between variables, correlation can also be used for prediction (in a statistical procedure called *regression* that is described in chapter 8). Further, correlation can be used in assessing reliability (for example, test-retest reliability; see chapter 13) and in assessing validity (for example, concurrent validity; see chapter 14).

GRAPHING CORRELATION

Correlation between two measures obtained from the same group of people can be shown graphically through the use of a scattergram. A **scattergram** (or a **scatterplot**) is a graphic presentation of a correlation between two variables (see figure 7.1). The two axes in the graph represent the two variables, and the points represent pairs of scores. Each point is located above a person's score on the horizontal axis (the *X* variable) and across from that person's score on the vertical axis (the *Y* variable). The *direction* of the correlation (positive or negative) and the *magnitude* of the correlation (ranging from –1.00 to +1.00) are depicted by a series of points.

Notice that the points on the scattergram in figure 7.1 create a pattern that goes from the bottom left upward to the top right. This is typical of a *positive* correlation, in which an *increase* in one variable is associated with an *increase* in the other variable.

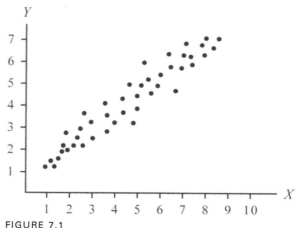

FIGURE 7.1

A scattergram showing a positive correlation between two variables, *X* and *Y*

The points on this scattergram cluster together to form a tight, diagonal pattern. This pattern is typical of a *high* (or *very high*) positive correlation.

To illustrate an example of the positive correlation displayed in the scattergram in figure 7.1, let's imagine the following fictitious study. In this study, the *X* variable is the amount of time (measured as hours per day) that students spend online, and the *Y* variable is the students' scores on a test that measures computer technological skills. According to figure 7.1, students who spend more time online score higher on the technology skills test. Conversely, those who spend a short amount of time online are less proficient in their technology skills. In real life, though, we rarely observe such high correlation between any two variables, especially those that measure behaviors or attitudes.

In a *negative* correlation, an *increase* in one variable is associated with a *decrease* in the other variable. For example, we can expect a negative correlation between days per year students are absent from school (*X* variable) and their grade point average (GPA) (*Y* variable). That is, as students are absent more and more days (an *increase* in *X*), their GPA falls lower and lower (a *decrease* in *Y*). The scattergram in figure 7.2 shows the hypothetical relationship between the two variables. Note that the direction of the points is from the top left downward toward the bottom right, indicating a negative correlation.

If you were to draw an imaginary line around the points on a scattergram, you would notice that as the correlation (positive or negative) gets higher, the points tend to cluster closer and form a clear pattern (figure 7.3). Thus, an inspection of the scattergram can indicate the approximate strength (or magnitude) of the correlation. For example, the scattergram in *part a*, where the points create a tight pattern, shows a higher correlation than that in *part b*, where the points are spread out wider.

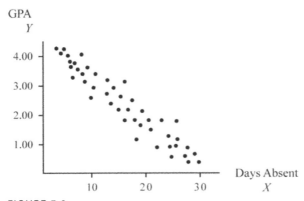

FIGURE 7.2
A scattergram showing a *negative* correlation between the number of days students are absent from school and their grade point average (GPA)

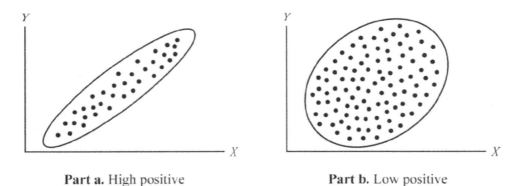

Part a. High positive   **Part b.** Low positive

FIGURE 7.3
A scattergram depicting two positive correlations: A high positive correlation (*part a*) and a lower positive correlation (*part b*)

When there is very low or no correlation between two variables, the scattergram contains points that do not form any clear pattern and are scattered widely (see figure 7.4).

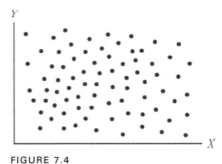

FIGURE 7.4
A scattergram showing no correlation between two variables, *X* and *Y*

Scattergrams can also be used to locate a specific pair of scores. For example, let's examine table 7.1 that lists scores on mathematics computation (variable *X*) and mathematics concepts (variable *Y*) for seven second-grade students. Figure 7.5 depicts the data in that table.

Table 7.1. Scores of Seven Students on Two Mathematical Tests

Student Number	Math Computation X	Math Concepts Y
A	18	20
B	17	15
C	11	12
D	19	18
E	13	12
F	15	16
G	17	18

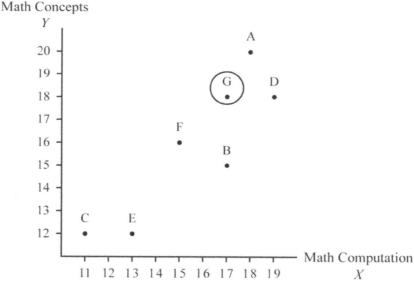

FIGURE 7.5
A scattergram of the correlation between math computation and math concepts (data in table 8.1)

Each point on the scattergram represents one student and corresponds to the scores listed in table 7.1. For example, let's find the point located at the top right-hand side of the scattergram in figure 7.5 that represents student G. We can draw a vertical line from that point down toward the *X* axis (math computation). Our line should intersect the axis at the score of 17. A horizontal line from point G toward the *Y* axis (math concepts) should intersect the axis at the score of 18. These scores—17 in math computation and 18 in math concepts—are indeed the same as those listed for student G in table 7.1.

A scattergram can help us identify scores that are noticeably different from the other scores. These scores, called **outliers,** can be easily spotted on a scattergram where they fall outside the range and pattern of the other points.[1] Figure 7.6 shows a scattergram with one outlier, located on the bottom right side.

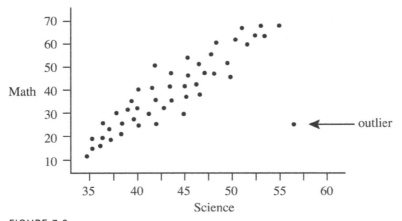

FIGURE 7.6
A scattergram showing the correlation between science and math scores with one outlier

PEARSON PRODUCT MOMENT

The most commonly used correlational procedure is the *Pearson product moment.* The **Pearson product-moment coefficient** (often referred to as **Pearson's r**) is named in honor of Karl Pearson (1857–1936), a British scientist who contributed a great deal to the development of statistics. Pearson was a student of Sir Francis Galton, who studied heredity. In 1896, Pearson developed the product-moment coefficient, which became quite popular within a short period of time. In order to use Pearson's correlation, the following requirements should be satisfied: (a) the scores are measured on an *interval* or *ratio* scale, and (b) the two variables to be correlated have a *linear* relationship (as opposed to *curvilinear* relationship).

To illustrate the difference between linear and curvilinear relationships, examine figure 7.7. *Part a* shows a *linear* relationship between height and weight, where the points form a pattern going in one direction. *Part b* shows a *curvilinear* relationship, where the age of individuals is correlated with their strength. Notice that the direction of the points is not consistent. In this example, the trend starts as a positive correlation and ends up as a negative correlation. For example, newborns are very weak and get stronger with age. They then reach an age when they are the strongest, and as they

---

1. HINT: Outliers are not unique to correlation. There may be outliers in any distribution for a variety of reasons. Researchers may wish to pay special attention to outliers and study them further.

age further, they become weaker. When Pearson's *r* is used with variables that have a curvilinear relationship, the resulting correlation is an *underestimate* of the true relationship between these variables.

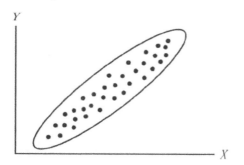 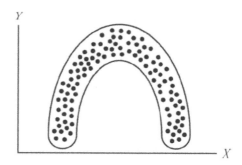

Part a. Linear Relationship   **Part b.** Curvilinear Relationship

FIGURE 7.7
Scattergrams showing a *linear* relationship between height and weight (*part a*) and a *curvilinear* relationship between age and strength (*part b*)

When observations on the variables to be correlated are rank-ordered, the statistic known as the *Spearman rank-order correlation* is used. The correlation coefficient is represented by r_s. This rank-order correlation coefficient is interpreted in the same way as the Pearson coefficient r.[2]

Interpreting the Correlation Coefficient

After obtaining the correlation coefficient, the next step is to evaluate and interpret it. It is important to remember that the *sign* of the correlation (negative or positive) is not indicative of the *strength* of the correlation and that a negative correlation is not something negative. What matters is the *absolute value* of the correlation. For instance, a negative correlation of $r = -.93$ indicates a stronger relationship than a positive correlation of $r = +.80$.

Table 7.2 lists guidelines for the interpretation of the strength of the correlation coefficients. These guidelines apply to both positive and negative correlation coefficients. There is no clear consensus among researchers as to the exact definition of each category, and the categories in the table overlap. Therefore, it is the researcher's decision how to define a certain coefficient. As an example, one researcher may describe a correlation coefficient of $r = .68$ as *high*, while another may define it as *moderate*. You can also use two categories to define a coefficient; for example, describing a correlation of $r = .40$ as *low-to-moderate* and a correlation of $r = .65$ as *moderate-to-high*.

---

2. HINT: The Greek letter ρ (rho) may also be used to indicate the rank-order correlation coefficient.

Table 7.2. An Interpretation of Correlation Coefficients

Correlation	Interpretation
.00–.20	Negligible to low (a correlation of .00 would be defined as "no correlation")
.20–.40	Low
.40–.60	Moderate
.60–.80	High/substantial
.80–1.00	Very high (a correlation of 1.00 would be defined as a "perfect correlation")

Another way to evaluate correlation coefficients is to divide the coefficients between $r = .00$ and $r = 1.00$ into three categories. Coefficients between $r = .00$ and $r = .33$ would be defined as low; coefficients between $r = .34$ and $r = .66$ would be considered moderate; and coefficients between $r = .67$ and $r = 1.00$ would be considered high. Again, two categories can be used to describe borderline coefficients. For instance, coefficients such as $r = .30$ or $r = .36$ can be described as low-to-moderate, and coefficients such as $r = .60$ or $r = .68$ can be described as moderate-to-high.

In describing and assessing correlations, it is important to consider the purpose of the study and the potential use of its results. For example, a correlation coefficient of $r = .60$ may be adequate for the purpose of *group* prediction but insufficient for *individual* prediction purposes.

The statistical significance level (p value) is often reported along with the coefficient itself. However, if the study involves the whole population and there is no attempt to generalize the results to other groups or settings, then the p value is not of importance. Rather, the obtained correlation coefficient is used to indicate the relationship between the variables.

The level of statistical significance is greatly affected by the sample size and might imply a high level of association between variables even when the correlation is low. For example, with a sample size of 350, even a low correlation, such as $r = .12$, is reported to be significant at $p < .05$, and a correlation of $r = .15$ is reported as significant at the $p < .01$ level. Hence, it is always a good idea to consider the *practical* significance of the correlation, along with its statistical significance.

Hypotheses for Correlation

The null hypothesis (H_O) states that there will be no correlation between the two variables being studied. That is, it predicts that in the population that is the focus of the investigation, the correlation coefficient is zero:

$$H_0 : r = 0$$

The alternative hypothesis (H_A) states that the population correlation is *not equal* to zero:

$$H_A : r \neq 0$$

After we obtain the correlation coefficient, we then consult a table of critical values. To use the table, we have to calculate the degrees of freedom (*df*) for the study. In correlation, the degrees of freedom are the number of *pairs* of scores minus 2. If the *obtained* correlation coefficient (*r*) *exceeds* the critical value, the null hypothesis is *rejected*. Rejecting the null hypothesis means that the chance that the correlation coefficient is 0 (*r* = 0) is very small and that *r* is large enough to be considered different from zero. When the obtained coefficient is *smaller* than the critical value, the null hypothesis is *retained*. We conclude that there is a high degree of likelihood that the correlation is not significantly different from 0.[3]

When your alternative hypothesis is stated as a *null* hypothesis (that is, it predicts no correlation between the two variables), you should use the **two-tailed test** to determine whether to *reject* or *retain* the null hypothesis. When your alternative hypothesis is *directional* and you predict that the correlation will be either positive or negative (but not zero), you should use the **one-tailed test.**[4]

When the null hypothesis is rejected, the level of significance (*p* level) is reported. This can be done using two approaches. There are researchers who choose to use the conventional "benchmarks" approach, where the level of significance (*p* level) is listed as *p* < .05, *p* < .02, or *p* < .01. Other researchers prefer to report the exact level of significance. Computer statistical packages usually print the exact *p* values, thus making this information readily available to researchers.

Computing and Assessing Pearson Correlation

To demonstrate the use of Pearson correlation, let's look at the scores of six students on two tests: reading vocabulary (*X*) and reading comprehension (*Y*) (table 7.3). Using a computer, we calculated that the correlation is *r* = .95. Note that the relative positions of students on the two tests are similar, which explains the high correlation. For example, student F scored low on both *X* and *Y*; and student D scored high on both *X* and *Y*.

Table 7.3. Scores of Six Students on Reading Vocabulary Test (*X*) and Reading Comprehension Test (*Y*)

Student	Reading Vocab. X	Reading Comp. Y
A	9	14
B	8	12
C	10	14
D	11	15
E	10	15
F	7	11

3. HINT: Although we present here the steps for calculating the degrees of freedom (*df*) and *p* value, computer statistical programs routinely calculate and provide you with the degrees of freedom and *p* values.

4. HINT: Certain statistical analysis software programs (such as SPSS) use the two-tailed test as default, but let you change it to one-tailed test.

After the correlation coefficient is calculated, the next step is to ascertain whether it is statistically significant. Table 7.4 lists the *critical values* of the Pearson *r* coefficient. In our example, the degrees of freedom (*df*) are 4 (the number of pairs minus 2).

Table 7.4. Partial Distribution of the Correlation Coefficient (Pearson's *r*) Critical Values (*df* = 4)

df	p level .10	p level .05	p level .02	p level .01
4	.729	.811	.882	.917

The critical values are .811 for a *p* level of .05, .882 for a *p* level of .02, and .917 for a *p* level of .01. These critical values can be listed as: r (.05, 4) = .811, r (.02, 4) = .882, and r (.01, 4) = .917. The .05, .02, and .01 listed inside the parentheses indicate the *p* level, and the number 4 indicates the degrees of freedom. Our calculated *r* value of .95 *exceeds* the critical values listed for .05, .02, and .01 levels of significance. Consequently, we *reject* the null hypothesis at *p* < .01 level. We conclude that a correlation coefficient of this magnitude (*r* = .95) could have occurred by chance alone less than 1 time in 100. (See chapter 2 for a discussion of the statistical hypothesis testing.)

FACTORS AFFECTING THE CORRELATION

The example in table 7.3 demonstrates the effect of the relative positions of individuals in their group on the correlation coefficient. The correlation is high if the following occurs: those who score high on *X* also score high on *Y*; those who score low on *X* also score low on *Y*; and those who score in the middle on *X* also score in the middle on *Y*. The actual scores on *X* and on *Y* do not have to be the same, only the relative position of scores in their group.

The *reliability* of the instruments used to collect data may also affect the correlation. The correlation coefficient may *underestimate* the true relationship between two variables if the measurement instruments used to obtain the scores are not reliable. (See chapter 13 for a discussion of reliability.)

Additionally, the correlation obtained may underestimate the real relationship between the variables if one or both variables have a *restricted range* (that is, low variance). To demonstrate an extreme case, suppose all of the students receive the same score on test *X*. (This may happen if the test is too easy.) If we try to correlate their scores on test *X* with their scores on another test, *Y*, we will get a correlation of zero (*r* = .00).

To illustrate this point, let's look at the scores of four students on two tests, *X* and *Y*, which are listed in table 7.5. Notice that all the students received the same score on *X* (*X* = 25). Figure 7.8 is a scattergram showing the same data. As you can see, the points that represent the students form a vertical line, as all the scores on the X variable have the same value. The students' scores on one variable are not related to their scores on the other variable.

Table 7.5. Scores of Four Students on Two Tests

Student	Test X	Test Y
A	25	15
B	25	18
C	25	17
D	25	14

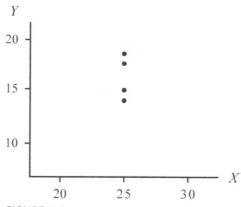

FIGURE 7.8

A scattergram of the data in table 7.5 showing r = .00, when all the scores on the X variable are the same (X = 25)

THE COEFFICIENT OF DETERMINATION AND EFFECT SIZE

When interpreting the correlation coefficient, researchers describe it in terms of its direction (positive or negative), magnitude (for example, low, moderate, or high), and its level of statistical significance (p value). Another index, called the *coefficient of determination,* or *shared variance,* describes how much individual difference in one variable is associated with individual difference in the other variable. The **coefficient of determination** (r^2) can be thought of as the percentage of the variability in one variable that can be attributed to differences in the scores on the other variable. This index is often used in prediction studies where one variable is used to predict another. (See chapter 8, which discusses prediction and regression.) The coefficient of determination is computed by squaring the correlation coefficient r. It can also be used as an index of the effect size. (See chapter 2 for a discussion of effect size.)

For example, suppose a college is using an admissions test that is given to all students who apply to this college. Assume further that based on results from past years the admissions office computes the correlation between past students' scores and their high school GPA and finds the correlation to be .50 ($r = .50$). The coefficient of determination is .25 or 25 percent ($r^2 = .50^2 = .25\%$). (See figure 7.9.) This coefficient of determination would allow the admissions office to assess the proportion of variability in one variable (admissions test score) that can be explained or determined by the other variable (high school GPA).

25%

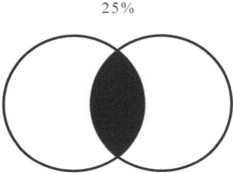

FIGURE 7.9
A graphic presentation of the relationship between two variables when $r = .50$ and the *coefficient of determination* (r^2) is 25%

When the correlation between two variables is $r = .90$, the coefficient of determination is 81 percent ($r^2 = .90^2 = .81 = 81\%$). Figure 7.10 illustrates the overlapping of the two variables.

81%

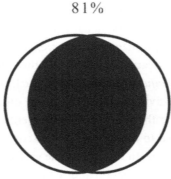

FIGURE 7.10
A graphic presentation of the relationship between two variables when $r = .90$ and the *coefficient of determination* (r^2) is 81%

INTERCORRELATION TABLES

At times, researchers are interested in correlating several variables with each other. Instead of embedding the results from such analyses in the text narrative, the results can be presented in an *intercorrelation table*. An **intercorrelation table** is used to display the correlations of several variables with each other. For example, let's say that the curriculum coordinator in a school district wants to take an in-depth look at the standardized test scores from one of the grade levels in the district and to see whether there are correlations between four subject areas: spelling, phonics, vocabulary, and language mechanics. The Total Battery test score is also included in the analysis. The curriculum coordinator displays the results in an intercorrelation table (see table 7.6).

Table 7.6.   Intercorrelations of Five Tests

	1 Spelling	2 Phonics	3 Vocab.	4 L. M.	5 Total
1. Spelling	1.00	.63	.45	.57	.82
2. Phonics	.63	1.00	.39	.68	.78
3. Vocabulary	.45	.39	1.00	.85	.86
4. Language Mechanics	.57	.68	.85	1.00	.91
5. Total Battery	.82	.78	.86	.91	1.00

As you can see, table 7.6 has two distinct features:

1. All of the correlations that are listed on the diagonal line in the center of the table (from top left to bottom right) are perfect ($r = 1.00$). The reason is obvious: these are all correlations of a variable with itself (for example, spelling and spelling, phonics and phonics).
2. If we divide the table into two triangles—top right and bottom left—we can see that the correlation coefficients that are recorded in the two triangles are the same and are the mirror image of each other.

Considering these two features, it is clear that table 7.6 contains *duplicate* information (in the two triangles) and *unnecessary* information (the correlations of 1.00 on the diagonal). Thus, the table can be reorganized to present the results more efficiently. Table 7.7 may look as if it lacks some information, but, in fact, it contains all the information needed.

Table 7.7.   A Revised Intercorrelation Table

	2 Phonics	3 Vocab.	4 L. M.	5 Total
1. Spelling	.63	.45	.57	.82
2. Phonics		.39	.68	.78
3. Vocabulary			.85	.86
4. Language Mechanics				.91
5. Total Battery				

You may find published reports where the lower left-hand side triangle is used to display the information, instead of the top right-hand side, as is the case in table 7.7. You may also find intercorrelation tables that include additional information about the measures that are being intercorrelated. For example, some of these tables may also list the means and standard deviations of the measures that were correlated.

In presenting an intercorrelation table, there are two ways to identify those correlation coefficients that are statistically significant. One approach is to display the significance levels with asterisks that indicate certain levels of significance (for example, $*$ $p < .05$ or $**$ $p < .01$). The other approach is to list the *exact* level of significance next to each correlation coefficient (for example, $p = .046$, or $p = .003$).

CORRELATION TABLES

Correlation tables differ from *intercorrelation* tables in the type of information they convey and in their layout. Both types of tables provide an efficient way to present a large number of correlation coefficients. In *correlation tables,* the variables listed in the rows and columns are different from each other, whereas in intercorrelation tables, the rows and columns list the same variables.

To illustrate the use of a correlation table, suppose we want to correlate the IQ levels of parents and their children (see table 7.8). The children in the study are divided into three groups: those with low, medium, and high IQ scores. In addition, IQ scores are available for the children's fathers and mothers.[5] Incidentally, note that in this example we do not have two scores for each participant, as was the case with other examples in this chapter (for example, spelling and phonics). Instead, in this example each child is paired with his/her parents.

Table 7.8. Correlations of IQ Scores of Mothers and Fathers with Their Daughters or Sons at Different IQ Levels

	Low IQ		Medium IQ		High IQ	
	Daughter	Son	Daughter	Son	Daughter	Son
Mother	$r = .56*$	$r = .62$	$r = .55**$	$r .44*$	$r = .48$	$r = .58*$
	$n = 15$	$n = 20$	$n = 22$	$n = 21$	$n = 16$	$n = 15$
Father	$r = .52$	$r = .45$	$r = .54**$	$r .48*$	$r = .49*$	$r = .50$
	$n = 14$	$n = 18$	$n = 23$	$n = 20$	$n = 17$	$n = 15$

$*$ $p < .05$
$**$ $p < .01$

In addition to the correlation coefficient (r) in table 7.8, the level of significance and the sample size (n) are also recorded in each cell. For instance, according to this table, we have correlational data for fifteen girls and their mothers ($n = 15$) in the low-IQ category. The correlation of the scores of these girls and their mothers is .56 ($r = .56$), significant at the .05 level ($*p < .05$).

---

5. HINT: As with any continuous variable that is divided into categories (such as high, medium, and low), the criterion used for creating the categories has to be logical.

SUMMARY

1. **Correlation** is defined as the relationship or association between two or more paired variables. The most common way to pair variables is to administer two measures to the same group of people and correlate their scores on the two measures.

2. The **correlation coefficient** indicates the strength (or degree) of correlation. The coefficient can range from $r = 1.00$ (perfect positive correlation) to $r = -1.00$ (perfect negative correlation). A coefficient of $r = 0.00$ indicates no correlation.

3. Correlation does not imply *causation*. Just because two variables correlate with each other does not mean that one caused the other.

4. Correlation is used to describe relationships between variables in prediction studies and in the assessment of reliability and validity.

5. A **scattergram** (or a **scatterplot**) is a graphic presentation of a correlation between two variables. The two axes in the graph represent the two variables, and the points represent pairs of scores.

6. The *direction* of the points on the scattergram and the degree to which they *cluster* indicate the strength of the correlation and whether the correlation is positive or negative. A scattergram can also show whether there are scores that are outliers.

7. In a *positive* correlation, an *increase* in one variable is associated with an *increase* in the other variable. In a *negative* correlation, an *increase* in one variable is associated with a *decrease* in the other variable.

8. The most commonly used correlation procedure is the *Pearson product moment*, whose coefficient is represented by the letter *r*. **Pearson's r** is used with data measured on an interval or a ratio scale when the variables to be correlated have a linear relationship.

9. Correlation coefficients can be described using words such as negligible, low, moderate, high, and very high. A combination of categories may also be used, such as moderate-to-high.

10. In describing and assessing correlations, it is important to consider the purpose of the study and the potential use of the results of the study. It is also important to consider the *practical* significance of the correlation and the effect size, along with its *statistical* significance.

11. The *null hypothesis* in correlation states that in the population the correlation coefficient is zero ($H_O = 0$), and the *alternative hypothesis* states that in the population the correlation is not equal to zero ($H_A \neq 0$).

12. The **two-tailed test** is used to determine whether to reject or retain the null hypothesis when the alternative hypothesis is stated as a *null* hypothesis and predicts no correlation between the two variables and the **one-tailed test** is used when the alternative hypothesis is directional.

13. The obtained correlation coefficient may be an *underestimate* of the real relationship between the variables if one or both variables have *low reliability* or if one or both variables have a *restricted range* (that is, low variance).

14. The **coefficient of determination** (r^2) (or *shared variance*) describes how much individual differences in one variable are associated with individual

differences in the other variable. This index is often used in prediction studies where one variable is used to predict another. The coefficient of determination is found by squaring the correlation coefficient. It can also be used as the *effect size* to assess the practical significance of the study's results.

15. Correlations between three or more variables are often presented in an intercorrelation or correlation table.

CHECK YOUR UNDERSTANDING

1. List at least one example each of *positive*, *negative*, and *low-to-zero* correlation. Explain your choices.
2. What is the potential effect of an *outlier score* on the correlation? Explain.
3. Suggest three hypotheses that can be tested using the Pearson correlation: *directional*, *nondirectional*, and *null*. Describe briefly a study that can be conducted to test each hypothesis.
4. When would you embed the results of a correlation study in the text and when would it be appropriate to use a table? Explain.

Prediction and Regression

In chapter 8 we extend the concept of correlation and show how it can be used in prediction. The statistical test that is introduced in this chapter is called *regression*. This is the process of using one variable to predict another when the two are correlated. It makes sense to expect that the higher the correlation between the variables, the more accurate the prediction. When one variable is used as a predictor, the statistical test is called *simple regression*; when two or more variables are used as predictors, *multiple regression* is used. We focus our discussion on simple regression and explain various terms and concepts. Regression can be depicted visually with the use of a scattergram, which has a *line of best fit* (regression line) drawn through the dots that create the scattergram. You will learn how to evaluate the line and how to decide if there are outliers; that is, scores that fall outside the pattern created by the rest of the scores. You can always expect a certain level of error when making a prediction, and we show you how to calculate and determine how accurate your prediction is going to be with the use of the *standard error of estimate*.

We include a numerical example to demonstrate and apply the concepts and terms used in this chapter. The statistical and practical significance of the results are also explained and discussed.

Prediction is based on the assumption that when two variables are correlated, we can use one to predict the other. The discussion in this chapter focuses on using prediction in educational settings. The examples demonstrate how educational and psychological instruments can be used in prediction.

PREDICTION AND REGRESSION

In our daily life, prediction is quite common. When we hear thunder and see lightning, we often predict they will be followed by rain. Similarly, we might predict the relationship between the day of the week and the expected crowd at the movie theater. In education, we also use prediction. For example, we might predict that a bright elementary school student will do well in high school or that a student who is having difficulties on the midterm examination is probably going to get a low grade on the final examination.

From our personal experience we know that our predictions do not always materialize, and people and events continue to surprise us. Sometimes rain does not follow thunder and lightning, and occasionally, bright young children drop out of high school. Nevertheless, knowing something about the relationship among the variables allows us to make a prediction that is better than a guess.

The technique used for prediction is called **regression**. When only one variable is used to predict another, the procedure is called **simple regression**, and when two or more variables are used as predictors, the procedure is called **multiple regression**.

The variable used as a *predictor* is the **independent variable**, and it is represented by the letter X. The predicted variable, represented by the letter Y, is called the **criterion variable**, or the *dependent variable*. For example, the Scholastic Aptitude Test (SAT) may be used as a predictor variable, and college freshman grade point average (GPA) may be the criterion variable.

The discussion that follows focuses on simple **linear regression**, where the predictor variable (X) and the criterion variable (Y) have a linear relationship.[1] Our numerical example demonstrates the computations involved in simple regression. Additionally, the concept of *multiple regression* is introduced briefly, without the use of a numerical example.

SIMPLE REGRESSION

After observing a high correlation between two variables, a researcher may want to use one variable to predict the other. For example, suppose a high school counselor has noticed that high-achieving students have higher academic self-concept compared with low-achieving students who have lower academic self-concept. The counselor may want to conduct a study to explore the idea of using a measure of academic self-concept (ASC) of high school students to predict their grade point average (GPA).

---

1. HINT: See chapter 7 for a discussion of the concept of linear relationship in correlation.

The counselor chooses a random sample of high school students to participate in the study. ASC scores and GPA are gathered for the students in the study and analyzed. If the ASC test is shown to be a good predictor of GPA, other educators may want to assess their students' academic self-concept at the beginning of the year and use this information in planning individualized instruction and coursework for their students.

Because correlation does not imply causation, we cannot conclude from the regression study that the students' academic self-concept has an effect on their GPA. Quite likely, both variables are related to ability; able students generally have higher academic self-concepts and they also get higher grades on their coursework. To ascertain whether ASC affects GPA, an *experimental* study should be conducted in which academic self-concept is manipulated.

The prediction of scores of a group of people on one variable from their scores on another variable can be done by using a *regression equation*. The equation is used to draw a line that is used for prediction. In order to develop the equation, we first need to have the predictor and criterion scores for a group of people. The members of this group should be *similar* to those whose criterion scores we would like to be able to predict in the future. Once the equation is available, it can be used to predict criterion (dependent) scores for a new group of people for whom only the predictor scores are available.

The regression equation can be used to draw a line through a scattergram of the two variables involved, designated as *X* and *Y*. This line is called the **regression line**, or the *line of best fit* (figure 8.1).

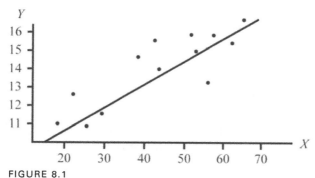

FIGURE 8.1
A scattergram showing the regression line

The position of the line is determined by the *slope* (the angle) and the *intercept* (the point where the line intersects the vertical axis, *Y*). The slope is represented by the letter *b*, and the intercept is represented by the letter *a*. The slope may also be referred to as the *coefficient*, and the intercept may be referred to as the *constant*.

Figure 8.2 illustrates four regression lines when the intercept is zero (that is, the regression line passes through the point where both axes are at zero). *Part a* shows that when $b = 0.25$, for every increase of 1 unit in X, there is an increase of 0.25 unit in Y; and when $b = 0.5$, for every increase of 1 unit in X, there is an increase of 0.5 units in Y (*part b*). When $b = 1$, for every increase of 1 unit in X, there is an increase of 1 unit in Y (*part c*); and when $b = 2$, for every increase of 1 unit in X, there is an increase of 2 units in Y (*part d*).

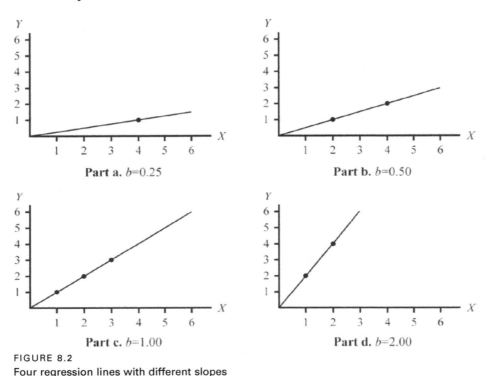

FIGURE 8.2
Four regression lines with different slopes

As figure 8.2 shows, you can see that the higher the value of b (the slope), the steeper the regression line; and the lower the value of b, the flatter the line.

The Standard Error of Estimate (S$_E$)

Unless the predictor and the criterion variables have a perfect correlation, any attempt to use X (the predictor) to predict Y' (the criterion) is likely to result in a certain degree of error. Consequently, for some individuals the Y' score (the predicted score) is an *overestimate* of their "true" Y score, while for others, the Y' score is an *underestimate* of their "true" Y score. The difference between the *actual* Y score and the *predicted* Y' score for each individual is called the *error score* (or *residual*). The standard deviation of the error scores, across all individuals, is called the **standard error of estimate (S$_E$).**[2]

More specifically, assuming a normal distribution of the error scores, the actual Y score would lie within $\pm 1S_E$ of the Y' score about 68 percent of the time, and within $\pm 2S_E$ about 95 percent of the time. The S$_E$ is calculated by using the scores from the group used to generate the regression equation. The formula for S$_E$ is:

$$S_E = S_Y \sqrt{1 - r^2}$$

Where S_E = Standard error of estimate
       S_Y = Standard deviation of the Y variable
       r^2 = Square of the correlation[3]

Holding S$_Y$ constant, S$_E$ *decreases* as *r increases*. Thus, the *higher* the correlation, the *lower* the S$_E$; and when the standard error of estimate is lower, the prediction is more accurate. The following formula demonstrates that when X and Y have a perfect correlation (r = 1.00), there is no error in prediction and the standard error of estimate (S$_E$) is zero:

$$S_E = S_Y \sqrt{1 - 1.00^2} = S_Y \sqrt{1 - 1.00} = S_Y \sqrt{0.00} = 0.00$$

By comparison, when r = 0.00 (there is no correlation), S$_E$ is equal to the SD of the Y variable (S$_Y$):

$$S_E = S_Y \sqrt{1 - 0.00} = S_Y \sqrt{1} = S_Y(1) = S_Y$$

An Example of Simple Regression

Ms. Wright, an eighth grade language arts teacher, wants to know whether she could use a practice test she constructed to predict the scores of her students on the state-mandated end-of-year language arts test. The teacher hypothesizes that the practice test administered at the beginning of the second semester is a good predictor of the state-mandated test. Thus, she might want to administer the practice test to her students, then use the test results to design early intervention and remediation programs for students who are expected to do poorly on the state-mandated test.

To ascertain whether the practice test is a good predictor of the state-mandated test, Ms. Wright uses the scores from the practice test (the *predictor*, or *independent* variable) and the scores from the state-mandated test (the *criterion*, or *dependent* variable) from her last year's students to generate the regression equation. Since the state-mandated

---

2. HINT: The standard error of estimate may also be represented by the symbol S$_{y \cdot x}$
3. HINT: You may recall that in chapter 7 we discussed the concept of r^2 (the coefficient of determination or shared variance) that refers to the proportion of the variability (or information) of Y that is contained in X.

language arts test is scored on a scale of 1 to 50, the teacher has designed her practice test to use the same scale.[4] In this computational example, the scores of ten students are used to demonstrate how to generate the regression equation (table 8.1).

Table 8.1. Scores of Ten Students on the Practice Test and on the State-Mandated Test

Student	Practice Test X	State Test Y
A	45	40
B	45	46
C	46	37
D	50	49
E	35	31
F	47	50
G	23	32
H	46	48
I	40	44
J	41	39
Mean	$\overline{X}$ = 41.80	$\overline{Y}$ = 41.60
Standard Deviation	S_X = 7.843	S_Y = 6.883

The teacher finds that the correlation between the two tests is $r = .764$.[5]

Next, the teacher computes the b coefficient, followed by the computation of the value of a.

$$b = r\frac{S_Y}{S_X} = (.764)\frac{6.883}{7.843} = (.764)(.878) = 0.671$$

$$a = \overline{Y} - b\overline{X} = 41.60 - (0.67)(41.80) = 41.60 - 28.03 = 13.57$$

After finding the values of b (the slope) and a (the intercept), they can be entered into the regression equation.

$$Y' = bX + a$$
$$Y' = 0.671(X) + 13.552$$

Now, after administering the practice test to her students, Ms. Wright can use the equation to predict their scores on the state-administered language arts test. For example, the teacher can predict that a student with a practice test (X) score of 30 is expected to have a score of 33.682 on the state test (Y'):

$$Y' = 0.671(30) + 13.552 = 20.13 + 13.552 = 33.682$$

4. HINT: In this example, we use numbers in a range of 1–50 for both the predictor (the practice test) and the criterion (the eighth grade end-of-year state-mandated test); however, the regression procedure allows for any range of scores to be used for the two variables.

5. HINT: The computations of the correlation coefficient are not included here. (See chapter 7 for a discussion of correlation.)

Of course, using this equation to predict the scores of new students on the state-mandated language arts test is predicated on the assumption that the new students taking the practice test are similar to those whose scores were used to derive the regression equation.

Using the equation above, we found that the standard error of estimate for the data in table 8.1 is 4.44. This means that for each student, on average, the teacher is likely to overestimate or underestimate the state-mandated language arts score by close to 4.5 points. For example, for students whose predicted Y' score is 42, about 68 percent of the time the actual Y score will lie within 4.44 above or below the Y' score (that is, between approximately 37.5 and 46.5). Even though this may look to you like a wide margin of error in prediction, remember that without having this information, it may have been even more difficult for the teacher to predict her students' scores!

Graphing the Regression Equation

As was mentioned before, the slope and the intercept in the regression equation can be used to draw a line through a scattergram that depicts the correlation of the two variables. Figure 8.3 shows a scattergram of the actual scores obtained on the practice test (X) and the scores obtained on the state-mandated test (Y) (table 8.1), with the regression line added.[6] All of the predicted Y' scores would lie on the regression line. Note that the regression line goes through the intersection of the means of the two variables (the mean of X is 41.80 and the mean of Y is 41.60).

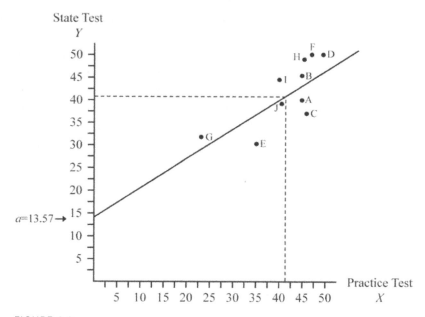

FIGURE 8.3
A regression line for predicting scores of ten students on the state-mandated test using the practice test scores as a predictor (see data in table 8.1)

6. HINT: Computer statistical programs can be used to draw the line.

As we can see, for several students, the Y' score is an overestimate (that is, the predicted Y' score is above the actual Y score); and for several students, the Y' score is an underestimate (that is, the predicted Y' score is below the actual Y score).

The Coefficient of Determination (r^2)

The *coefficient of determination* (r^2) can be used to describe the relationship between the variables.[7] In our example, the language arts teacher used her own practice test to predict her students' scores on the end-of-year state-mandated language arts test. The teacher found that the correlation between the two tests was $r = .764$. As you may recall, to find the coefficient of determination we need to square the correlation (r^2). With $r = .764$, the coefficient of determination is $r^2 = .764^2 = .584$ (or 58%). This coefficient means that about 58 percent of the variation in performance on the state test (Y) can be accounted for by individual differences in performance on the practice test (X); 42 percent of the variation is due to other factors. As was discussed in chapter 2, the coefficient of determination (r^2) can also be used as an index of *effect size*.

MULTIPLE REGRESSION

The *multiple regression* procedure is utilized when two or more variables are used to predict one criterion variable. The combined correlation of the predictor variables with the criterion variable is called **multiple correlation**, represented by the symbol **R**. For example, scores on a kindergarten readiness test, combined with teacher assessment scores, may be used to predict first graders' scores on a standardized achievement test. A researcher is likely to consider using several variables as predictors when there is no single variable that has a high correlation with the criterion so as to serve as a satisfactory single predictor. In such cases, additional predictor variables may be used in order to predict the criterion variable more accurately. In our example, combining the two predictor variables (kindergarten readiness test and teacher assessment) is likely to predict the criterion variable (first grade standardized achievement test) more accurately, compared with using only one of the two predictors.

The regression equation in multiple regression is an extension of the equation for simple regression. In addition to the intercept (a), the equation contains a regression coefficient (b) for each of the predictor variables (X_1 and X_2). With two predictors, the equation is:

$$Y' = b_1 X_1 + b_2 X_2 + a$$

Where Y' = Predicted Y score
$\quad\quad b_1$ = Slope (coefficient) of predictor X_1
$\quad\quad X_1$ = Score on independent variable (predictor) X_1
$\quad\quad b_2$ = Slope (coefficient) of predictor X_2
$\quad\quad X_2$ = Score on independent variable (predictor) X_2
$\quad\quad a$ = Intercept (constant)

---

7. HINT: This concept was discussed in chapter 7 (Correlation).

In multiple regression, the *coefficient of determination* is represented by R^2, which is similar to r^2 in simple regression. Just like r^2, the coefficient of determination in multiple regression can range from 0 to 1.00. R^2 indicates the proportion of the variation in Y that can be accounted for by the variation of the *combined* predictor variables.

R^2 is greater when the predictor variables have a low correlation with each other than when the predictor variables correlate highly with each other. To illustrate this point, let's look at figure 8.4.

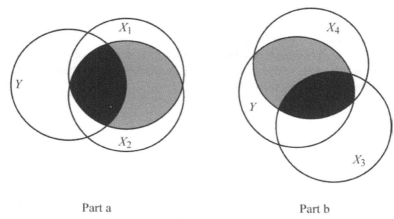

Part a   Part b

FIGURE 8.4
Two graphs showing different levels of correlation between the two predictors, and between the predictors and the criterion: the two predictors: X_1 and X_2, correlate highly with each other (*part a*); and the two predictors, X_3 and X_4, have a low correlation with each other (*part b*)

Part a shows two predictor variables, X_1 and X_2, which correlate highly with the criterion variable Y. In addition, the two predictors also correlate highly with each other. The high correlation between the two predictor variables is evidenced by the fact that they overlap a great deal. Adding a second predictor (X_2) to the first predictor (X_1) has not significantly increased R^2, the amount of variation in Y (the criterion) that can be accounted for by the predictors. That is, adding a second predictor does not account for a much greater proportion of the criterion Y.

As we can tell, in *part a* there is a large, white area on the left side that represents the proportion in variable Y that is not accounted for by the two predictors.

Part b shows two predictor variables, X_3 and X_4, and a criterion variable Y. Each of the two predictors has a *high* correlation with the criterion variable and a *low* correlation with each other (X_3 and X_4 overlap very little). In *part b*, the much smaller white area in variable Y shows that the two combined predictor variables cover more of the criterion variable and can account for more of the variation in Y.

SUMMARY

1. **Regression** is a technique used for prediction. The variable used as a predictor is called the **independent variable** and is represented by the letter X. The predicted variable is called the **criterion** or the *dependent* variable and is represented by the letter Y.

2. Regression is based on the assumption that the predictor (or predictors) and the criterion variable correlate with each other. The higher the correlation, the more accurate the prediction.

3. When one variable is used to predict another variable, the procedure is called **simple regression**. When two or more variables are used as predictors, the procedure is called **multiple regression**.

4. **Linear regression** is used when the predictor variable (X) and the criterion variable (Y) have a linear relationship.

5. The regression equation is used to predict Y scores for a given group of individuals for whom the X scores are available. The predicted Y score is represented by Y'. The equation is:

$$Y' = bX + a$$

6. The *regression equation* can be used to draw a line. The position of the line is determined by its *slope*, represented by the letter b, and by its *intercept* (the point where the regression line intersects the vertical axis), represented by the letter a. The slope may also be identified as the *coefficient*, and the intercept may be identified as the *constant*.

7. The characteristics of the sample used to derive the regression equation should be similar to those of the sample for which we want to predict future scores.

8. When inspecting regression lines, we can see that the higher the value of b, the steeper the line, and the lower the value of b, the flatter the line.

9. Unless the predictor and the criterion variables have a perfect correlation, any attempt to use the predictor variable (X) to predict the criterion variable (Y') is likely to result in a certain degree of error. The *error score* (or the *residual*) is the difference between actual Y and predicted Y' scores.

10. The standard deviation of the error scores across all individuals is called the **standard error of estimate** (S_E). S_E indicates how much, on average, Y' scores *overestimate* or *underestimate* the actual Y scores.

11. As the correlation between variables increases, S_E decreases, thereby making the prediction more accurate.

12. The *coefficient of determination*, r^2, can be used to describe the relationship between the two variables. It describes the amount of variation in the scores on the criterion variable Y that can be accounted for by individual differences on X, the predictor variable.

13. The *multiple regression* procedure is employed when two or more variables are used to predict one criterion variable. A researcher is likely to consider using two or more variables as predictors when there is no single variable

that has a high correlation with the criterion so as to serve as a satisfactory predictor. In such cases, additional predictor variables may be used in order to predict the criterion variable more accurately.

14. The regression equation for multiple regression is an extension of the equation for simple regression. It includes one intercept (a, the constant); and a series of slopes (b the regression coefficient), one for each of the predictor variables. With two predictors (X_1 and X_2), the multiple regression prediction equation is:

$$Y' = b_1 X_1 + b_2 X_2 + a$$

15. The combined correlation of the predictor variables with the criterion variable is called **multiple correlation R**.

16. R^2 (which is similar to r^2 in simple regression) indicates the proportion of the variation in Y (the criterion variable) that can be accounted for by the variation of the combined predictor variables.

CHECK YOUR UNDERSTANDING

1. Give an example of a study where *simple regression* might be used in prediction, including the selection of *independent* (*predictor*) and *criterion* (*dependent*) variables. Explain your choice.

2. Why, in a *simple regression* study, is it better to use *independent* and *dependent* variables that have a high correlation? Explain.

3. How can the *regression line* be used to illustrate *simple* regression? ·

4. Give an example of a prediction study where *multiple regression* might be used. Explain your choice.

IV

COMPARING GROUP MEANS

9

t Test

The previous two chapters discussed approaches to data analysis when there are two or more variables that are *correlated*. Chapter 9 looks at a different research inquiry: *comparing two means* to ascertain which mean is significantly higher. The statistical test used for this type of analysis is the *t test,* and the statistic that is computed is called a *t value*. In this chapter, you will learn about three research situations in which the *t* test can be used to analyze the data and compare the means from: (a) *two independent groups* (for example, experimental and control); (b) *two paired samples* (for example, pretest and posttest scores); and (c) *a sample and a population* (for example, comparing a mean of a sample to the mean of the population).

Examples with numerical data are included in the chapter to illustrate each of the three types of *t* tests. As you will see in these examples, after the *t* value is computed you have to decide whether it is *statistically* significant and whether your research hypothesis was confirmed. We also show you how to evaluate the *practical* significance of your findings and how to draw conclusions about your study.

Many researchers are concerned with research questions that relate to the comparison of groups to each other. Their studies are conducted to determine whether differences between groups are statistically significant or whether they could have occurred simply by chance. When two means are being compared with each other, the statistic used is a *t* **test**. The *t* test is based on the *t distribution* that was developed in 1908 by W. S. Gosset, who worked for a brewery in Dublin. Since employees were not allowed to publish in journals, Gosset used the pseudonym "Student" in an article he sent to a journal, and the *t* distribution became known as "Student's *t* distribution."

USING THE *t* TEST IN EDUCATION

There are a variety of research situations where a *t* test can be used to analyze the data. For example, research studies may be conducted to compare an experimental group to a control group, boys to girls, or pretest to posttest scores. The numbers used for the comparison are the means of the two groups. The scores used to compute the means should be measured on an interval or ratio scale and be derived from the same measure (for example, the same test). As we analyze research data, we should keep in mind that small differences are expected even among members of the same group. These differences may occur due to sampling error and are considered chance differences.

You might ask, "How can we distinguish between differences due to sampling error and 'real' differences? At what point do we say that the difference is too large to be attributed to sampling error, and that it probably indicates a real difference?" Unfortunately, there are no standards or cutoff scores. After obtaining the means, we cannot simply "eyeball" them and determine whether they are similar or different. A difference of two points between means may be defined as statistically significant in some cases, but not in others. The group means and variances, in addition to the sample size, all play a role in determining whether the difference between the means is a "real" difference.

HYPOTHESES FOR *t* TESTS

Predictions of outcomes in studies that are using the *t* test reflect what the researcher hypothesizes about the nature of the differences between the means. The alternative hypothesis (that is, the research hypothesis), represented by the symbol H_A or H_1, predicts whether there would be a statistically significant difference between the two means being compared.[1] For example, the alternative hypothesis may be a *directional hypothesis* that predicts that teachers who collaborate and share instructional resources and approaches (experimental group) are more likely to value teacher action research compared with similar teachers who do not collaborate (control group).

$$H_A: \text{Mean}_E > \text{Mean}_C$$

---

1. HINT: See chapter 2 for a full explanation of alternative (research) and null hypotheses.

Where H_A = Research hypothesis (the alternative hypothesis)
   Mean$_E$ = Mean of the experimental group
   Mean$_C$ = Mean of the control group

A *nondirectional hypothesis* in a *t* test predicts that there would be a difference between the two means, but the direction of the outcome is not specified. For example, we may predict that there are differences in attitudes toward involvement in community volunteering projects between junior high boys and girls, but due to inconclusive results in previous studies, we are unable to predict which of the two groups will have a more positive attitude. The nondirectional hypothesis is:

$$H_A : \text{Mean}_1 \neq \text{Mean}_2$$

Occasionally, the research hypothesis is not stated as directional or nondirectional, but in a null form. That is, we predict that there will be *no difference* between the means. This is not very common in educational research, but in cases where the research hypothesis is stated as null, that hypothesis is considered nondirectional.

The *null hypothesis* (H_O) in the *t* test states that any observed difference between the means is too small to indicate a real difference between them and that such difference is probably due to sampling error. In other words, the null hypothesis always predicts no difference between the means.[2] For example, we may predict that there would be no difference in the number of hours spent on doing homework as reported by boys and girls in two neighboring middle schools. In symbols, the null hypothesis is:

$$H_O: \text{Mean}_1 = \text{Mean}_2$$

$$\text{Or: } H_O: \text{Mean}_1 - \text{Mean}_2 = 0$$

The null hypothesis is submitted to a statistical test and based on the results of this statistical test, we decide whether to retain or reject the null hypothesis. Since the null hypothesis always predicts no difference, there is no need to formally state it when the research hypotheses are presented. But when the research (alternative) hypothesis predicts no difference between the means, then it should be stated and included among the other study's research hypotheses.

In order to calculate the *t* test value, a score should be obtained for each person or case. The scores are then used to calculate the *t* value. After calculating a *t* value, the next step is to consult the table of critical values for the *t* distribution in order to determine the level of significance (*p* value) of the obtained *t* value. In order to use the table, the researcher needs to know whether to use the critical values listed under the *one-tailed* or *two-tailed* test. If your research hypothesis is *directional* and you predict

---

2. HINT: This is true when we use the *t* test. You may remember that in correlational studies, the null hypothesis always predicts no *relationship* between the variables being correlated (that is, it predicts a correlation that does not differ significantly from zero; see chapter 7).

which mean will be higher, use the *one-tailed test*. If your research hypothesis is *non-directional* and you predict a difference between the means but do not specify which mean will be higher, use a *two-tailed test*. If your hypothesis is stated as *null* and you predict no difference between the means beyond what might be expected purely by chance, use the *two-tailed test*. When in doubt, use the two-tailed test, which is considered more conservative.[3]

Nowadays, with computer statistical software packages readily available, there is no need to compute the *p* value by hand. The computer program will provide the appropriate *p* value and the *t* value, along with the two means and standard deviations. Nonetheless, you will still need to know whether to use a one-tailed or two-tailed test.

A *t* test is used to compare two means in three different situations:

1. *t* **test for independent samples**. The two groups whose means are being compared are independent of each other. A typical example is a comparison of experimental and control groups.
2. *t* **test for paired samples** (also called a *t* test for *dependent, matched,* or *correlated* samples). The two means represent two sets of scores that are paired. A typical example is a comparison of pretest and posttest scores obtained from one group of people.
3. *t* **test for a single sample**. This *t* test is used when the mean of a sample is compared to the mean of a population. For example, we may use the Graduate Record Examination (GRE) scores of psychology graduate students (the sample) to test whether they are significantly different from the overall mean GRE score in the university (the population).

In using the *t* test, it is assumed that the scores of those in the groups being studied are normally distributed and that the groups were randomly selected from their respective populations. In reality, in studies that are conducted in educational and behavioral sciences, it is quite difficult to satisfy these requirements. Nonetheless, empirical studies have demonstrated that we can use the *t* test even if the assumptions are not fully met.

t TEST FOR INDEPENDENT SAMPLES

The *t* test for independent samples is used extensively in experimental designs and in causal comparative (*ex post facto*) designs when means from two groups are being compared. There are several assumptions underlying this test:

1. The groups are independent of each other.
2. A person (or case) may appear in only one group.

---

3. HINT: A conservative test or a conservative decision generally reduces the chance of making a Type I error. (See chapter 2.)

3. The two groups come from two populations whose variances are approximately the same. This assumption is called the *assumption of the homogeneity of variances*. We compare the two variances to determine if there is a statistically significant difference between them. When the two groups are approximately the same size, there is no need to test for the homogeneity of variances.

To test for the assumption of the homogeneity of the variances, we divide the larger variance by the smaller variance and obtain a ratio, called the *F value* (or *F ratio*). A *test for the equality of variances*, such as the *Levene's test*, is used to test the significance of the *F* value. An *F* value that is *not* statistically significant ($p > .05$) indicates that the assumption for the homogeneity of variances was *not* violated and, hence, *equal variances* can be assumed. A *significant F* value ($p < .05$) indicates that the assumption for the homogeneity of variances was violated and the *t* test statistical results should then be adjusted for the *unequal variances*.

It is unlikely that you will have to do the computations by hand in order to test for the homogeneity of variances. Most statistical software packages (such as SPSS) include the level of significance of the *F* value in their report of results for the independent-samples *t* test, as well as the adjustment of the *t* value where needed.

The *t* test is considered a *robust* statistic. Thus, even if the assumption of the homogeneity of variance is not fully met, the researcher can probably still use the test to analyze the data. As a general rule, *it is desirable to have similar group sizes, especially when the groups are small*.[4]

An Example of a *t* Test for Independent Samples

A new test preparation company, called Bright Future (BF), wants to convince high school students studying for the American College Testing (ACT) test that enrolling in their test preparation course would significantly improve the students' ACT scores. BF selects twenty students at random and assigns ten to an experimental group and ten to a control group.[5] The students in the experimental group participate in the test preparation course conducted by BF. At the conclusion of the course, both groups of students take the ACT test, which was given to high school students the previous year. BF conducts a *t* test for independent samples to compare the scores of Group 1 (Experimental E) to those of Group 2 (Control C). The scores of the ten students in each of the two groups, their groups' means, standard deviations, and variances, are listed in table 9.1

---

4. HINT: As in several other statistical tests, researchers usually try to have a group size of at least thirty. Larger samples are more stable and require a smaller *t* value (compared with smaller samples) to reject the null hypothesis.

5. HINT: Although in real-life studies researchers try to have larger sample sizes, in this chapter (as well as in other chapters) we are using small sample sizes to simplify the computations in the examples given.

Table 9.1. ACT Scores of Experimental Group (n=10) and Control Group (n=10)

	(Experimental) Group 1	(Control) Group 2
	26	19
	27	24
	21	18
	31	23
	21	22
	25	24
	29	32
	32	29
	34	15
	23	20
Mean	$\bar{X}_1 = 26.90$	$\bar{X}_2 = 22.60$
SD	$S = 4.56$	$S = 5.08$
Variance	$S_1^2 = 20.77$	$S_2^2 = 25.83$
Sample size	$n_1 = 10$	$n_2 = 10$

The study's research (alternative) hypothesis (H_A) is directional and can be described as:

$$H_A : \mu_E > \mu_C$$

Note that μ, the symbol for the population mean, is used in writing hypotheses. Remember that although we may conduct our studies using *samples*, we are testing hypotheses about *populations*. The null hypothesis states that there is no significant difference between the two means. A study is then designed to test the null hypothesis and to decide whether it is tenable. The null hypothesis is:

$$H_0 : \mu_E = \mu_C$$

The *t* value is computed using this formula:

$$t = \frac{\bar{X}_1 - \bar{X}_2}{\sqrt{\frac{(n_1-1)S_1^2 + (n_2-1)S_2^2}{n_1+n_2-2} \left(\frac{1}{n_1} + \frac{1}{n_2} \right)}}$$

Where X_1 = Mean of Group 1
$\quad\quad X_2$ = Mean of Group 2
$\quad\quad S_1$ = Variance of Group 1
$\quad\quad S_2$ = Variance of Group 2
$\quad\quad n_1$ = Number of people in Group 1
$\quad\quad n_2$ = Number of people in Group 2

As we can see, in addition to the difference between the two means (the numerator), the formula also includes the two sample sizes and the two variances (the denominator). This can explain why we cannot simply look at the difference between the two means and decide whether that difference is statistically significant. The number of scores in each group and the variability of these scores also play a role in the *t* test calculations. In other words, the difference between the means is viewed in relation to the number of scores and their spread. When the spread is small (a low variance), even a small difference between the means may lead to results that are considered statistically significant. With a larger spread (a higher variance), a relatively large difference between the means may be required in order to obtain results that are considered statistically significant.

After finding the *t* value, it is then compared to appropriate critical values in the *t test table of critical values*. When the obtained *t* test value *exceeds* its appropriate critical value, the null hypothesis is *rejected*. This allows us to conclude that there is a high level of probability that the difference between the means is notably greater than zero and that a difference of this magnitude is unlikely to have occurred by chance alone. When the obtained *t* test *does not exceed* the critical value, the null hypothesis is *retained*.

Following are the computations of the *t* value for students in experimental and control groups who participated in a study conducted by BF. In our example, using the groups' means, variances, and sample sizes, we find the *t* value to be 1.992.

$$t = \frac{26.90 - 22.60}{\sqrt{\dfrac{(10-1)(20.77)+(10-1)(25.83)}{10+10-2}\left(\dfrac{1}{10}+\dfrac{1}{10}\right)}} =$$

$$\frac{4.30}{\sqrt{\dfrac{(186.93+232.47)}{18}}(0.10+0.10)} = \frac{4.30}{\sqrt{\dfrac{419.40}{18}}(0.20)} =$$

$$= \frac{4.30}{\sqrt{(23.30)(0.20)}} = \frac{4.30}{\sqrt{4.66}} = \frac{4.30}{2.159} = 1.992$$

Next, we need to determine the appropriate critical value and decide whether to reject or retain the null hypothesis. We calculate the degrees of freedom to be 18 ($df =$ (10 + 10 − 2 = 18), and because our hypothesis was directional, we use the one-tailed test. Table 9.2 shows a section from the table of critical values for t tests.

Table 9.2.   Partial Distribution of the t Test Critical Values ($df = 18$)

	Level of Significance for One-Tailed Test			
p values	.10	.05	.025	.01
df = 18	1.330	1.734	2.101	2.552

As is the convention, we start by examining the critical value under a p value of .05. In our case, it is 1.734 for a one-tailed test. (This is listed as: t_{crit}(.05, 18) = 1.734, with .05 showing the p level and 18 indicating the df.) We find the critical value by locating p of .05 under the one-tailed test, and df of 18. Our obtained t value of 1.992 *exceeds* the corresponding critical value of 1.734; therefore, we move to the right to the next column and check the critical value at $p = .025$, which is 2.101 (t_{crit}(.025, 18) = 2.101). Our obtained value of 1.992 *does not* exceed this critical value. Thus, we report our results to be significant at the $p < .05$ level, which is the last critical value that we did exceed.

Our decision is to reject the null hypothesis. Such a large difference between the two groups could have occurred by chance alone less than 5 percent of the time.[6] The hypothesis stated by BF Company is confirmed: students who participated in the test-taking course scored significantly higher on the practice form of the ACT than did the students in the control group. Based on the results of this study, we can be at least 95 percent confident that the course offered by BF is indeed helpful to students similar to those who participated in the study.

As was discussed in chapter 2, reporting the *statistical* significance of an experimental study should be followed by an evaluation of its *practical* significance. In addition to inspecting and evaluating the difference between the means of the two groups, we can use the index of effect size (ES) to evaluate the practical significance of our study and the effectiveness of the intervention. As you recall, the effect size is calculated using this formula:

$$ES = \frac{\text{Mean}_{EXP} - \text{Mean}_{CONT}}{\text{SD}_{CONT}}$$

The numerator is the difference between the means of the experimental and control groups, and the denominator is the standard deviation of the control group. Entering our means and standard deviations yields an effect size of .85.

---

6. HINT: You should remember, though, that we are making our statistical decision in terms of *probability*, not *certainty*. Rejecting the null hypothesis at $p < .05$ means that there is a possible error associated with this decision.

$$ES = \frac{26.90 - 22.60}{5.082} = \frac{4.30}{5.082} = .85$$

Our ES of .85, is considered high. This ES confirms that the difference of 4.30 points between the two groups is *practically* significant, in addition to being *statistically* significant (at $p < .05$). This information seems to support the statistical significance that indicated that the likelihood of getting such a large difference between the two groups purely by chance is less than 5 percent. We can conclude that participating in the BF test-taking preparation course is effective and can bring about a statistically and practically significant increase in students' ACT scores. Of course, this was a study with a very small sample size, and BF may need to repeat the study with larger samples to really convince students and their parents!

t TEST FOR PAIRED SAMPLES

A *t* test for paired samples is used when the two means being compared come from two sets of scores that are related to each other. It is used, for example, in experimental research to measure the effect of an intervention by comparing the posttest to the pretest scores. The most important requirement for conducting this *t* test is that the two sets of scores are *paired*. In studies using a pretest-posttest design, it is easy to see how the scores are paired: they belong to the same individuals. It is assumed that the two sets of scores are normally distributed and that the samples were randomly selected.

To compute the paired-samples *t* test, we need to first find for each person the difference (D) between the two scores (for example, between pretest and posttest) and sum up those differences (ΣD). Usually, the lower scores (for example, pretest) are subtracted from the higher scores (for example, posttest) so D values are positive. We also need to compute the sum of the squares of the differences (ΣD^2). The *t* value is computed using this formula:

$$t = \frac{\sum D}{\sqrt{\dfrac{n\left(\sum D^2\right) - \left(\sum D\right)^2}{n-1}}}$$

Where ΣD = Sum of the difference scores (D)

ΣD^2 = Sum of the squared differences (D^2)

n = Number of pairs of scores

The example that follows demonstrates the computation of a paired-samples *t* test. To simplify the computations, we use the scores of eight students only. Of course, in conducting real-life research, it is recommended that larger samples (thirty or more) be used.

An Example of a *t* Test for Paired Samples

Research has documented the potential effect of students' positive self-concept on their self-perceptions, attitudes toward self, schoolwork, and general development. A special program is developed by school psychologists and primary grade teachers to enhance the self-concept of young, school-age children. The program is implemented in Sunny Bay School with four groups of first and second grade students. The intervention lasts six weeks and involves various activities in the class and at home. The instrument used to assess the effectiveness of the program is comprised of 40 pictures, and scores can range from 0 to 40. The program coordinator conducts a series of workshops to train five graduate students to administer the instrument. All of the children in the program are tested before the start of the program and then again one week after the end of the program. A *t* test for paired samples is used to test the hypothesis that students' self-concept would improve significantly on the posttest, as compared with their pretest scores. The research hypothesis is:

$$H_A: \text{Mean}_{POST} > \text{Mean}_{PRE}$$

Table 9.3 shows the numerical data we used to compute the *t* value for the eight students selected at random from the program participants. The table shows the pretest and posttest scores for each student, as well as the means on the pretest and posttest. The third column in the table shows the difference between each pair of scores (D) and is created by subtracting the pretest from the posttest for each participant. The gain scores are then squared and recorded in the fourth column (D^2). The scores in these last two columns are added up to create ΣD and ΣD^2, respectively. These values are used in the computation of the *t* value.

Table 9.3. Self-Concept Pretest and Posttest Scores of Eight Students

Pretest X^1	Posttest X^2	Posttest–Pretest D	(Posttest–Pretest)^2 D^2
30	31	+1	1
31	32	+1	1
34	35	+1	1
32	40	+8	64
32	32	0	0
30	31	+1	1
33	35	+2	4
34	37	+3	9
$\bar{X}_1 = 32.00$	$\bar{X}_2 = 34.13$	$\Sigma D = 17$	$\Sigma D^2 = 81$

$$t = \frac{\sum D}{\sqrt{\dfrac{n\sum D^2 - \left(\sum D\right)^2}{n-1}}} = \frac{17}{\sqrt{\dfrac{(8)(81) - (17)^2}{8-1}}} =$$

$$= \frac{17}{\sqrt{\dfrac{359}{7}}} = \frac{17}{\sqrt{51.29}} = \frac{17}{7.16} = 2.37$$

Our computations show that the obtained *t* value is 2.37. This obtained *t* value is then compared to the values in the abbreviated table of critical values of the *t* distribution (table 9.4). The obtained *t* value of 2.37 exceeds the critical values for one-tailed test under $p = .05$ and the critical value of $p = .025$. The obtained *t* value *does not* exceed the critical value under $p < .01$, which is 2.998. Therefore, we reject the null hypothesis that states that there is no difference between the pretest and the posttest scores and report our results as significant at the $p < .025$ level. The likelihood that these results were obtained purely by chance is less than 2.5 percent. We confirm the research hypothesis that predicted that the posttest mean score would be significantly higher than the pretest mean score. According to this study, the self-concept enhancement program was effective in increasing the self-concept of first and second grade students.

Table 9.4. Partial Distribution of the *t* Test Critical Values (*df* = 7)

	Level of Significance for One-Tailed Test					
p values	.10	.05	.025	.01	.005	.0005
df = 7	1.415	1.895	2.365	2.998	3.499	5.405

Although our data seem to indicate that the intervention to increase the self-concept of the primary grade students was effective, those conducting the research or those reading about it should still decide for themselves whether the intervention is worthwhile. The question to be asked is whether an increase of 2.13 points (out of 40 possible points on the scale) is worth the investment of time, money, and effort.

t TEST FOR A SINGLE SAMPLE

Occasionally, a researcher is interested in comparing a single group (a sample) to a larger group (a population). For example, a high school teacher of a freshmen accelerated English class may want to confirm that the students in that class had obtained higher scores on an English placement test compared with their peers. In order to carry out this kind of a study the researcher *must* know prior to the start of the study the mean value of the population. In this example, the mean score of the population is the overall mean of the scores of *all* freshmen on the English placement test.

An Example of a *t* Test for a Single Sample

A kindergarten teacher in a school commented to her colleague that the students in her class this year seem to be less bright than those she had in the past. Her colleague disagrees with her. To test whether this year's first graders are really different from those in previous years, they conduct a *t* test for a single sample. The scores used are from the Wechsler Preschool and Primary Scale of Intelligence, Third Edition (WPPSI-IV), which is given every year to all kindergarten students in the district. In this example, we consider the district to be the population to which we compare the mean of the current kindergarten class. Although the mean IQ score of the population at large is 100 (μ =100), this district's mean IQ score is 110 (μ = 110), and this mean is used in the analysis. The research hypothesis is stated as a null hypothesis and predicts that there is no statistically significant difference in the mean IQ score of this year's kindergarten students (the sample) and the mean IQ score of all kindergarten students in the district (the population) that were gathered and recorded over the last three years.

$$H_A : \text{Mean}_{\text{CLASS}} = \text{Mean}_{\text{DISTRICT}}$$

The formula for the *t* test of a single sample is:

$$t = \frac{\overline{X} - \mu}{S_{\overline{X}}}$$

Where $\overline{X}$ = Sample mean
μ = Population mean
$S_{\overline{x}}$ = Standard error of the mean

To find $S_{\overline{x}}$ we use this formula:

$$S_{\overline{X}} = \frac{S}{\sqrt{n}}$$

Where $S_{\overline{x}}$ = Standard error of the mean
S = Sample standard deviation
n = Number of individuals in the sample

In order to test their hypothesis, the two teachers randomly select IQ scores of ten students from this year's kindergarten class. These IQ scores are listed in table 9.5, followed by the computation of the *t* value.

**Table 9.5. IQ Scores of
10 Students**

Scores	
115	118
135	113
105	98
107	120
112	99
$\sum X =$	1122
$\overline{X} =$	112.20
S (SD) =	10.94

$$S_{\bar{x}} = \frac{S}{\sqrt{n}} = \frac{10.94}{\sqrt{10}} = \frac{10.94}{3.16} = 3.46$$

$$t = \frac{\overline{X} - \mu}{S_{\bar{x}}} = \frac{112.20 - 110}{3.46} = \frac{2.20}{3.46} = 0.64$$

The degrees of freedom in this study are 9 (number of students in the sample minus one). We examine the *p* values for a two-tailed *t* test because our hypothesis is stated as a null hypothesis. Table 9.6 shows the critical values for *df* = 9.

Table 9.6. Partial Distribution of the *t* Test Critical Values (*df* = 9)

p values	Level of Significance for One-Tailed Test					
	.20	.10	.05	.02	.01	.001
df = 9	1.383	1.833	2.262	2.821	3.250	4.781

The obtained *t* value of 0.64 *does not* exceed the critical value under *p* = .05, which is 2.262. Thus, the null hypothesis is retained. Based on these results, the two teachers conclude that there is no significant difference between this year's kindergarten class and the "typical" kindergarten students in the district. In fact, the mean IQ score of this year's kindergarten students (mean = 112.20) is actually slightly higher than the mean score of the district (mean = 110). The research hypothesis that was stated in a null form (that is, predicting no difference between the two means) is confirmed.

SUMMARY

1. The **t test** is used to compare two means to determine whether the difference between them is statistically significant.
2. The *t* test requires data measured on an *interval* or *ratio* scale.

3. A *directional* hypothesis in the *t* test predicts which of the two means is going to be higher (H_A: $Mean_1$ > $Mean_2$)

4. A *nondirectional* hypothesis in a *t* test predicts a difference between the two means but does not specify which mean will be higher (H_A: $Mean_1$ ≠ $Mean_2$).

5. The *null* hypothesis in a *t* test states that any observed difference between the means is too small to indicate a real difference between them and that such a difference is probably due to sampling error. In other words, the null hypothesis always predicts no difference between the means beyond what might happen purely by chance (H_O: $Mean_1$ = $Mean_2$).

6. When using the table of *critical values*, *directional hypotheses* are tested using the *one-tailed test*, and *nondirectional hypotheses* are tested using the *two-tailed test*. When in doubt, use the two-tailed test, which is more conservative.

7. The *t* test can be used to compare means from: (a) two independent groups, (b) two paired groups, and (c) a single sample and a population.

8. In using the *t* test, it is assumed that the scores of those in the groups being studied are normally distributed and that the groups were randomly selected from their respective populations. Although in studies conducted in educational and behavioral sciences it is sometimes difficult to satisfy these requirements, empirical studies have demonstrated that we can use the *t* test even if these assumptions are not fully met.

9. The **t test for independent samples** is used when the two groups or sets of scores whose means are being compared are independent of each other. When conducting the *t* test for independent samples, it is assumed that the two groups being compared come from two populations whose variances are approximately the same. This assumption is called *the assumption of the homogeneity of variances*. We can compare the two variances to check whether the difference between them is statistically significant.

10. The *t* test is considered a *robust* statistic; therefore, even if the assumption about the homogeneity of the variance is not met, the researcher can still safely use the test to analyze the data. As a general rule, it is desirable to have similar group sizes, especially when the groups are small.

11. In studies that compare the means of experimental and control groups, the *effect size* (ES) may also be calculated, in addition to the *t* value.

12. The **t test for paired samples** (also called a *t* test for *dependent*, *matched*, or *correlated* samples) is used when the means come from two sets of paired scores.

13. The **t test for a single sample** is used when the mean of a sample is compared to the mean of a population. In order to carry out this kind of a study the researcher *must* know prior to the start of the study the mean value of the population.

CHECK YOUR UNDERSTANDING

1. What are the differences between *correlation* and the *t* test? Provide examples and explain at least two differences.
2. Suggest three hypotheses that can be tested using a *t test*: *directional*, *nondirectional*, and *null*. Describe briefly a study that can be conducted to test each hypothesis.
3. How is *effect size* used in *t test* studies where *experimental* and *control* groups are used? Explain.

Analysis of Variance

Chapter 10 presents the statistical procedure called *analysis of variance* (abbreviated as ANOVA). This is a procedure used to test for significant differences between means when there are *two or more means*. There are several types of ANOVA tests; in this book we introduce you to the two that are the most common. The test statistic that is computed in ANOVA is called the *F ratio*. One-way ANOVA can be thought of as an extension of the *t* test for independent samples. However, when there are multiple means to compare, ANOVA also can compare all means simultaneously, whereas to do so using the *t* test, you would need to conduct a series of comparisons, two means at a time. The concept of one-way ANOVA is explained in detail with the use of graphs. The numerical example in the chapter walks you step-by-step through the data analysis and interpretation. As you will see, when the results of the one-way ANOVA are statistically significant, we conduct a *post hoc pairwise comparison* to determine which means are different from each other.

When there are two or more independent variables, we use a *factorial ANOVA*. In this chapter, we introduce you to a test called *two-way ANOVA*. You may wish to use this test when you have two independent variables and you want to test for an *interaction* between the two. Again, a numerical example is used to illustrate this statistical test.

In conducting research, we often want to compare multiple means generated from several groups. The **analysis of variance** (**ANOVA**) test, which was developed by R. A. Fisher in the early 1920s, is used to compare the means of two or more independent samples to ascertain whether the differences between the means are statistically significant.[1] In this chapter, we explain the concept of ANOVA, followed by specific examples that further illustrate its advantages and uses.

UNDERSTANDING ANOVA

To better understand the importance of ANOVA, suppose, for example, that we want to compare five groups. If a *t* test is used, we have to repeat it ten times to compare all the means to each other; we have to compare the mean from group 1 with the means from groups 2, 3, 4, and 5; and the mean from group 2 with the means from groups 3, 4, and 5; and so on. Every time we do a *t* test, there is a certain level of error that is associated with our decision to reject the null hypothesis; the error is compounded as we repeat the test over and over. The main risk is that we may make a Type I error, that is, reject the null hypothesis when in fact it is true and should be retained (see chapter 2). By comparison, when we use ANOVA to compare the five means simultaneously, the error level can be kept at the .05 level (5 percent). In addition to keeping the error level at a minimum, performing one ANOVA procedure is more efficient than doing a series of *t* tests.

In ANOVA, the *independent variable* is the categorical variable that defines the groups that are compared (for example, instructional methods, grade level, or marital status). The *dependent variable* is the measured variable whose means are being compared (for example, language arts test scores, level of job satisfaction, or test anxiety).

There are several assumptions for ANOVA: (a) the groups are independent of each other, (b) the dependent variable is measured on an interval or ratio scale, (c) the dependent variable is normally distributed in the population, (d) the scores are random samples from their respective populations, and (e) the variances of the populations from which the samples were drawn are equal (the assumption of the *homogeneity of variances*). The first two assumptions (*a* and *b*) must always be satisfied. Assumptions *c* and *d* are often difficult to satisfy in education and behavioral sciences. Nonetheless, even if we cannot determine that random sampling was used, we can generally satisfy the requirement that the samples are not biased. ANOVA is considered a *robust* statistic that can stand some violation of the third and fourth assumptions, and empirical studies show that there are no serious negative consequences if these assumptions are not fully met. The last assumption (*e*) can be tested using special tests, such as the *F* test, in which the largest variance is divided by the smallest variance.

The test statistic in ANOVA is called the *F* statistic. The **F ratio** (or **F value**) is computed by dividing two variance estimates by each other.[2] If the *F* ratio is statistically

---

1. HINT: Although ANOVA can be used with two or more groups, most researchers use the independent samples *t* test when the study involves two independent groups, whereas ANOVA is used when there are three or more independent groups in the study. The results obtained from each approach would provide similar results and lead to the same conclusions.

2. HINT: Even though the ANOVA test is designed to compare means, the samples' *variances* and *variance estimates* are used in the computation of the *F* ratio.

significant (that is, if $p < .05$) and if there are three or more groups in the study, then a pairwise **post hoc comparison** is done to assess which two means are significantly different from each other.[3]

When one independent variable is used, the test is called a **one-way analysis of variance (one-way ANOVA)**. To illustrate, let's look at a week-long study to test the effect of instructional methods on middle school students' understanding of one of the standards from the National Science Education Standards, which states: "If more than one force acts on an object along a straight line then the forces will reinforce or cancel each other depending on their direction and magnitude; unbalanced forces will cause changes in the speed or direction of an object's motion." In the study, three instructional methods are used with three groups of seventh grade classes: teacher-led expository approach, guided inquiry approach, and free discovery approach. At the end of the week, students are tested to demonstrate their understanding of the standard. The F ratio would be used to assess whether there are significant differences in the mean scores of the three groups of students.

The ANOVA test can be applied to studies with more than one independent variable. For example, we can study the relationship between *two* independent variables and a dependent variable. When there are two independent variables, the design is called a **two-way analysis of variance (two-way ANOVA)**. In general, when two or more independent variables are studied in ANOVA, the design is called a *factorial analysis of variance*.

Let's go back to our example of the three instructional methods and seventh grade students' science scores. The three instructional methods were teacher-led expository approach, guided inquiry approach, and free discovery approach. Based on prior research and our own experience, suppose we believe that the students' gender also makes a difference and that science scores of seventh grade boys and girls would differ depending on the instructional method used by their teacher. A two-way ANOVA can be used to explore the effect of the two independent variables (instructional method and gender) on the dependent variable (science scores). In addition, using the two-way ANOVA would allow us to study the interactions among all variables. In the factorial ANOVA test, the **interaction** refers to a situation where one or more levels of the independent variable have a different effect on the dependent variable when combined with another independent variable. For example, we may find that boys score higher in science when their teacher uses one of the three instructional approaches, whereas girls' science scores are higher when their teacher uses another instructional approach.

An independent variable must have at least two *levels* (or conditions). For example, the variable of gender has two levels (female and male) and the variable of the seasons

---

3. HINT: An example of the pairwise comparison is discussed later in this chapter in the section titled *Post Hoc Multiple Comparisons*.

of the year has four levels (fall, winter, spring, and summer). To further explain the concept of levels of independent variables, let's look at the following examples of two-way and three-way ANOVA tests.

Suppose we conduct a study of high school students to investigate the relationship between two independent variables, gender and socioeconomic status (SES), and their effect on students' attitudes toward school (the dependent variable). The variable of gender has two levels (female and male), and the variable of SES in this study has three levels (low, middle, and high). The design of the study is indicated as a *2 × 3 factorial ANOVA* (or a *two-way ANOVA*). Assume we want to add a third variable to our study, such as the level of education of the students' parents. In our study, the variable of parents' education would have three levels or categories. We would assign a code of 1–3 to each student depending on the parents' level of education. A code of 1 would be assigned to students if their parents *did not go* to college; a code of 2 would be assigned to students if *at least one* of their parents has had a minimum of two years of college education; a code of 3 would be assigned to students if both parents have had a minimum of two years of college education. The design of our study would be: 2 × 3 × 3 ANOVA.

ONE-WAY ANOVA

Conceptualizing the One-Way ANOVA

ANOVA is used to study three types of variability that are called the **sum of squares** (abbreviated as **SS**). They are:

1. *Within-groups sum of squares (SS$_W$)*, which is the variability within the groups.
2. *Between-groups sum of squares groups (SS$_B$)*, which is the average variability of the means of the groups around the total mean. (The total mean is the mean of all the scores, combined.) (It may also be called *among-groups sum of squares*, abbreviated as SS$_A$).
3. *Total sum of squares (SS$_T$)*, the variability of all the scores around the total mean.[4]

The *total* sum of squares is equal to the combined *within*-groups sum of squares and the *between*-groups sum of squares:

$$SS_T = SS_W + SS_B$$

Figure 10.1 illustrates the different sum of squares and shows that the sum of squares *within* (SS$_W$) plus the total sum *between* (SS$_B$) equal the *total* sum of squares (SS$_T$). In this figure, X_1 is the score of an individual in group 1; $\bar{x}_1$ is the mean of group 1; and $\bar{x}_T$ is the total mean.

---

4. HINT: Think of it as combining the scores from all the groups to create one large group and computing the variability of this group around the total mean.

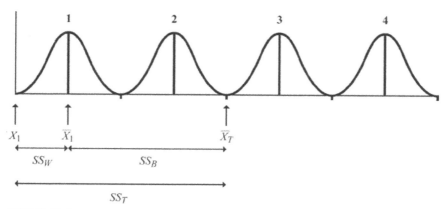

FIGURE 10.1

A graph showing the three sums of squares: SS_W, SS_B, and SS_T

After finding the sums of squares, the next step is to compute the variance esti-mates. The variance estimates are called the **mean squares**, abbreviated as **MS**. The mean squares are found by dividing the sum of squares (SS) by the appropriate degrees of freedom. This process is similar to the computation of a variance for a single sample where we divided the sum of the squared distances of scores around their means (the sum of squared deviations) by $n − 1$ (see chapter 5). In ANOVA, the degrees of freedom are used in place of $N − 1$ as the denominator when com-puting the variance.

The degrees of freedom for within (dfW) are found by subtracting the number of groups in the study (that is, K) from the total number of individuals in the study ($N − K$). The degrees of freedom for between (dfB) are found by subtracting 1 from the total number of groups ($K − 1$). The degrees of freedom associated with the total variance estimate (dfT) are the total number of scores minus 1 ($N − 1$). The dfT is equal to the combined degrees of freedom for the within mean square (dfW) and between mean squares (dfB). For example, let's say we have three groups in our study with twenty people in each group. The total number of people in the study is 60 ($3 × 20 = 60$). In this example, $K = 3$, $n = 20$, and $N = 60$. We calculate the degrees of freedom as follows:

$dfW = N − K = 60 − 3 = 57$

$dfB = K − 1 = 3 − 1 = 2$

$dfT = N − 1 = 60 − 1 = 59$

To compute the F ratio, we need only two variance estimates, MS_W and MS_B. As a result, there is no need to compute the total mean square (MS_T). The formulas for computing MS_W and MS_B are:

$$MS_W = \frac{SS_W}{N-K} \qquad MS_B = \frac{SS_B}{K-1}$$

The MS_W (also called the *error term*) can be thought of as the average variance to be expected in any normally distributed group. The MS_W serves as the denominator in the computation of the F ratio. To compute the F ratio, the *between*-group mean square (MS_B) is divided by the within-group mean square (MS_W).

$$F = \frac{MS_B}{MS_W}$$

MS_B, the numerator, increases as the differences between the group means increase; thus, greater differences between the means also result in a higher F ratio. Additionally, since the denominator is the within-group mean square, when the groups are more homogeneous and have lower variances, the MS_W tends to be smaller and the F ratio is likely to be higher. Two figures are presented to illustrate the role of group means (figure 10.2) and variances (figure 10.3) in the computations of the F ratio in ANOVA.

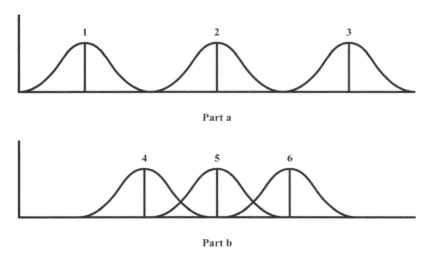

Part a

Part b

FIGURE 10.2
A graph showing three groups with different means and similar variances (*part a*), and a graph showing three groups with more similar means and similar variances (*part b*)

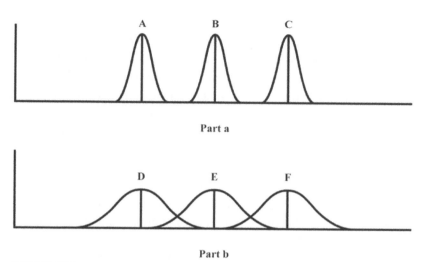

Part a

Part b

FIGURE 10.3
A graph showing three distributions with small variances (*part a*), and a graph showing three distributions with the same means as the groups in *part a* but with higher variances (*part b*)

Part a and *part b* in figure 10.2 show the distributions of scores of several groups. The variances of the three groups in *part a* (groups 1, 2, and 3) are about the same as the variances of the three groups in *part b* (groups 4, 5, and 6). By comparison, the means of the three groups in *part a* are farther apart from each other, compared with the means of the three groups in *part b*. If asked to predict which part of figure 10.2 would yield a higher F ratio, we would probably choose *part a*, where the three groups do not overlap and the means are quite different from each other. By contrast, the means of groups 4, 5, and 6 are closer to each other, and the three distributions overlap.

Next, let's turn our attention to figure 10.3, where two sets of distributions are presented in *part a* (groups A, B, and C) and in *part b* (groups D, E, and F). Notice that the three groups in *part a* have the same *means* as the three groups in *part b*, but the *variances* in both parts are different. The variances of the three groups depicted in *part a* are very low (that is, the groups are homogeneous with regard to the characteristic being measured). By comparison, the variances of the three groups depicted in *part b* are high, with a wide spread of scores in each group. If we were to compute an F ratio for the two sets of scores, we can predict that the F ratio computed for the three groups in *part a* would probably be high and statistically significant, whereas the F ratio computed for the three groups in *part b* would probably be lower and not statistically significant. This prediction is based on the knowledge that when the group variances are low (as is the case in *part a*), MS_w (the denominator in the F ratio computation) is low and we are more likely to get a high F ratio. The variances in *part b* are higher than those in *part a*, and we can expect a higher MS_W and a lower F ratio.

Once the F ratio is obtained, we have to decide whether to retain or reject the null hypothesis. *Retaining* the null hypothesis means that the sample means are not significantly different from each other beyond what might be expected purely by chance, and we consider them as coming from the same population. *Rejecting* the null hypothesis means that at least two sample means differ significantly from each other.

In studies where the null hypothesis is rejected, the next step is to conduct a post hoc comparison, in which all possible pairs of means are compared in order to find out which pair(s) of means differ(s). When the researcher predicts which means are expected to differ *before* starting the investigation, a method of *a priori* (or *planned*) *comparisons* is used to test this prediction. *A priori* comparisons are appropriate when the researcher has a sound basis for predicting the outcomes before starting the study, while post hoc comparisons are appropriate in exploratory studies or when no specific prediction is made prior to the start of the study.

Hypotheses for a One-Way ANOVA

A one-way ANOVA tests the null hypothesis (H_O) that states that all the groups represent populations that have the same means. When there are three means, the *null* hypothesis is:[5]

$$H_0 : \mu_1 = \mu_2 = \mu_3$$

The *alternative* hypothesis, H_A (also called the *research hypothesis*), states that there is a statistically significant difference between at least two means. When there are three groups, the alternative hypothesis is:

$$H_A : \mu_1 \neq \mu_2 \quad \frac{and}{or} \quad \mu_1 \neq \mu_3 \quad \frac{and}{or} \quad \mu_2 \neq \mu_3$$

The ANOVA Summary Table

The results of the ANOVA computations are often displayed in a summary table (see table 10.1). This table lists the sum of squares (SS), degrees of freedom (df), mean squares (MS), F ratio (F), and the level of significance (p level). The general format of the ANOVA summary table is presented in table 10.1.

Instead of being displayed in a summary table, the results may also be incorporated into the text. The information in the text includes the F ratio, the degrees of freedom for the numerator (dfB) and the degrees of freedom for the denominator (dfW). The information is listed as F(dfB,dfW). The text most likely will also include the level of statistical significance (p level).

---

5. HINT: Although *samples* are studied, as with other statistical tests, we are interested in the *populations* that are represented by these samples. Consequently, in the null and alternative hypotheses, μ (the Greek letter *mu*) is used to represent the population means.

Table 10.1. The General Format of the One-Way ANOVA Summary Table

Source	SS	df	MS	F	p
Between	SS_B	K-1	MS_B	F-ratio	<.05>
Within	SS_W	N-K	MS_W		
Total	SS_T	N-1			

Further Interpretation of the *F* Ratio

Figure 10.4, *parts a, b,* and *c,* represent three hypothetical samples and their *F* ratios. Each hypothetical example shows a study where three groups were compared.

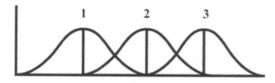

Part a. *F*-ratio is probably not significant (retain null)

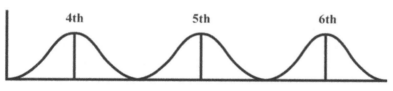

Part b. *F*-ratio is probably significant (reject null)

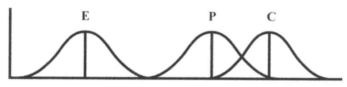

Part c. *F*-ratio is probably significant (reject null)

FIGURE 10.4

Three sets of distributions showing different *F* ratios: A nonsignificant *F* ratio (*part a*); a significant *F* ratio (*part b* and *part c*)

Part a depicts language arts test scores of seventh grade students from three neighboring school districts. Note that the distributions of the three groups of students overlap, and the means are not very different from each other. The *F* ratio comparing the students from the three districts is probably small and not statistically significant. *Part b* shows mean scores from a fifth grade mathematics test given to fourth, fifth, and sixth grade classes. As is expected, the fourth graders scored the lowest, and the sixth graders scored the highest. The F ratio comparing the fourth and sixth graders is statistically significant. The figure also shows clear differences between the fourth and fifth graders and between the fifth and sixth graders. For that reason, the ANOVA test that compares these groups to each other is likely to yield an F ratio that is statistically significant. *Part c* shows mean scores on an aggression scale, given to three groups after an intervention designed to decrease aggression. The three groups are: control (C), placebo (P), and experimental (E). As we can see, after the intervention the experimental group had the lowest aggression mean score, followed by the placebo group, while the control group scored the highest. The difference between the experimental group and the placebo group may be statistically significant, and, quite likely, there is a statistically significant difference between the experimental and control groups. The difference between the placebo and the control groups is probably not statistically significant. We can speculate that in this hypothetical example, the F ratio is probably large enough to lead us to reject the null hypothesis.

An Example of a One-Way ANOVA

Professor Learner, the statistics course instructor at Midwestern State University, wants to test four instructional methods for teaching statistics. Students who signed up to take her statistics course are assigned at random to four sections: Section 1 is taught online; section 2 is taught using lectures; section 3 is taught using independent study; and section 4 is taught using a combination of lectures, group work, online work, and watching videos on the topic. In this study, the instructional methods are the independent variable. Students in all four sections have to take five quizzes and a comprehensive final examination (the dependent variable). The scores on each quiz can range from 1 to 15. With four sections, the null hypothesis (H_O) and the alternative hypothesis (H_A) are:

$$H_0 : \mu_1 = \mu_2 = \mu_3 = \mu_4$$
$$H_A : \mu_i \neq \mu_j$$

The subscripts i and j can represent any two of the four sections. In other words, the alternative hypothesis predicts that there will be a significant difference between at least two of the four means. The null hypothesis predicts that there will be no significant differences between the section means beyond what might happen purely by chance, due to some sampling error.

To illustrate the computations of ANOVA, we chose at random the scores of five students from each section on one of the quizzes. These quiz scores are the dependent variable. Of course, if this was a real study, we would have used a much larger sample size!

Table 10.2 lists the scores of the twenty randomly selected students. The sample sizes and means (n and $\bar{x}$) are listed for each section and for the total group (N and

$\bar{x}_T$). As we can see, there are differences between the means of the four groups ($\bar{x}_1 = 13.8$, $\bar{x}_2 = 13.2$, $\bar{x}_3 = 11.2$, and $\bar{x}_4 = 14.6$). The question is whether these differences are statistically significant or are due to chance; ANOVA can help us answer this question.

(Note that our numerical example does not include the computational steps because you are likely to use a computer program to do the calculations for you.)

Table 10.2.   Raw Scores and Summary Scores of Four Groups on a Statistics Quiz

Online Section 1	Lectures Section 2	Independent Study Section 3	Combined Section 4	TOTAL
14	14	11	15	
15	13	10	14	
13	11	11	15	
13	13	14	14	
14	15	10	15	
$n_1 = 5$	$n_2 = 5$	$n_3 = 5$	$n_4 = 5$	$N_T = 20$
$\bar{X}_1 = 13.80$	$\bar{X}_2 = 13.20$	$\bar{X}_3 = 11.20$	$\bar{X}_4 = 14.60$	$\bar{X}_T = 13.20$

The numerical results are displayed in an ANOVA summary table (table 10.3). The table lists the three sources of variability (SS), the three degrees of freedom (*df*), the two mean squares (MS), the *F* ratio, and the *p* value.

Table 10.3.   One-Way ANOVA Summary Table for the Data in Table 10.2

Source	SS	df	MS	F	p
Between	31.6	3	10.53	7.14	< .01
Within	23.5	16	1.48		
Total	55.2	19			

Our conclusion is that the teaching method *does make a difference* in the students' test scores. As indicated in table 10.2, students in section 4, where the teaching method was a combination of lectures, group work, online work, and videos obtained the highest quiz scores. The second highest mean quiz score was obtained by students in section 1 (Online), followed by section 2 (Lectures). The students in section 3 (Independent Study) obtained the lowest mean score on the quiz.

Since our *F* ratio was significant at the $p < .01$ level, our next step is to conduct a *post hoc comparison* to find out which means are significantly different from each other. In our example, we use the *Tukey method* for the post hoc comparisons.

Post Hoc Multiple Comparisons

The *Tukey method* of post hoc multiple comparisons is also called the *honestly significant difference (HSD)*. The group means are compared to each other in a pairwise analysis. Because the results in table 10.3 indicate that there was a statistically significant difference between at least two means, we can compare the four group means to each other, one pair at a time. This analysis is usually done by computer using statistical software packages. In our example, we found that there was a statistically significant difference at $p < .05$ between the Online method ($\bar{x}_1{-} = 13.80$) and the Independent Study method ($\bar{x}_3 = 11.20$); and at $p < .01$ between the Combined method ($\bar{x}_4 = 14.60$) and the Online method ($\bar{x}_1{-} = 13.80$). No other means are significantly different from each other. Figure 10.5 depicts the four groups and their means.

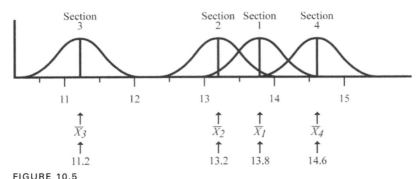

FIGURE 10.5
A graph illustrating the data in table 10.2

The likelihood that the decision to reject the null hypothesis is wrong is low, and Professor Learner can be quite confident that her choice of instructional method in teaching statistics affects the students' quiz scores. The professor can examine the results in order to decide which instructional methods to use in teaching her statistics course in the future.

TWO-WAY ANOVA

Conceptualizing the Two-Way ANOVA

The *two-way* ANOVA test is designed to study the relationship between *two or more* independent variables and a dependent variable. One advantage of the two-way ANOVA is that it can reveal an *interaction* between the two independent variables. This interaction may not be apparent when a series of one-way ANOVA tests is conducted. To illustrate this point, let's look at the example that was used to demonstrate the computation of a one-way ANOVA.

Four different instructional methods were tested with four sections of students enrolled in a college statistics course. Students who registered to take the course were assigned at random to one of the four sections. The means on a quiz administered to students in all four sections were compared to each other using a one-way ANOVA.

Suppose we want to further divide the students in each group by their major in college, by ability level, or by gender. It is possible to conduct another ANOVA test to compare, for example, the quiz scores of psychology students in all four instructional methods to the scores of their classmates who major in history. We can run two separate one-way ANOVA tests; one to compare the four methods and one to compare the two majors. Nonetheless, instead of running these two tests, we can do a single two-way ANOVA test. The two-way ANOVA would allow us to compare simultaneously the method effect, the college major effect, and the possible effect of the interaction between the method and the major on the students' quiz scores. For example, psychology students may score higher using one instructional method, whereas history students may do better using another method.

The total variation in a two-way ANOVA is partitioned into two main sources: the *within-groups* variation and the *between-groups* variation. The between-group variation is further partitioned into three components: (a) the variation among the *row means*, (b) the variation among the *column means*, and (c) the variation due to the *interaction*.

Four mean squares (MS) are computed in a two-way ANOVA. Two are computed for the two *main effects* (the independent variables), one is computed for the *interaction*, and one for the *within*. Then, using the mean squares within (MS_W) as the denominator, three F ratios are computed. These F ratios are found by dividing each of the three mean squares (MS_{Row}, MS_{Column}, and $MS_{Row} X _{Column}$) by MS_W. As was the case with a one-way ANOVA, a summary table is used to display the two-way ANOVA summary information. The table includes the sum of squares, degrees of freedom, mean squares, F ratios, and p levels.

Hypotheses for the Two-Way ANOVA

A two-way ANOVA is conducted to test three null hypotheses about the effect of each of the two independent variables on the dependent variable and about the interaction between the two independent variables. The two independent variables (or *factors*) are referred to as the *row* variable and the *column* variable.

To test the three null hypotheses, three F ratios are calculated in a two-way ANOVA. The three null hypotheses are:

- H_O ($_{Row}$): There are no statistically significant differences among population means on the dependent measure for the first factor (the *row factor*).
- H_O ($_{Column}$): There are no statistically significant differences among the population means on the dependent measure for the second factor (the *column factor*).
- H_O ($_{Interaction}$): In the population, there is no statistically significant interaction between factor 1 and factor 2 (the *row × column interaction*).

Graphing the Interaction

It is often helpful to further study the interaction by graphing it. To create *the interaction graph*, the mean scores on the dependent variable are marked on the vertical axis. Lines are then used to connect the means of the groups.

Suppose, for example, that we want to conduct a study to investigate two methods designed to increase the attention span of third and sixth grade students using two behavior modification methods. In this study, students' grade level is one independent variable; the second independent variable is the behavior modification method. Half of the students in grade 3 and half of the students in grade 6 are taught using method 1. The other half in each class is taught using method 2. The dependent variable is the students' attention span. Figure 10.6 shows two possible outcomes of the study: the interaction is significant and the lines intersect (*part a*); and the interaction is not significant and the lines are parallel (*part b*).

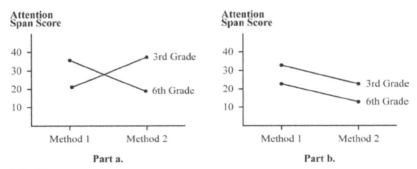

FIGURE 10.6
A graph showing a statistically significant interaction (*part a*); and a graph showing an interaction that is not statistically significant (*part b*)

Part a in figure 10.6 shows that there was an interaction effect between the behavior modification method and student grade level. We can conclude that method 1 was more effective with the sixth graders, and method 2 was more effective with the third graders regarding their attention span. *Part b* shows no interaction effect. Method 1 was more effective for both the third graders and the sixth graders, and method 2 was less effective for both grade levels. In addition, the sixth graders who were taught using method 1 and the sixth graders who were taught using method 2 scored lower than the third graders who were taught using either method 1 or method 2.

There are two types of significant interactions: (a) *disordinal*, where the lines intersect (*part a* in figure 10.6); and (b) *ordinal*, where the lines do not intersect (see figure 10.7). Therefore, an interaction may be significant even if the two lines do not intersect, as long as they are *not parallel*.

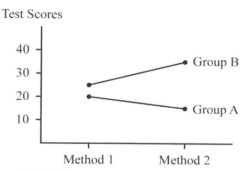

Test Scores

FIGURE 10.7
A graph showing an ordinal significant interaction representing the data in table 10.4

To illustrate an ordinal interaction, let's look at the data in table 10.4. In this hypothetical example, four sixth grade classes in two middle schools are taught history of ancient civilizations using two different teaching methods. One *independent* variable is the students' school (group A and group B), and the other *independent* variable is the instructional method (method 1 and method 2). The *dependent* variable is the students' scores on the final ancient civilizations test. One sixth grade class in each of the two schools is taught using method 1, and the other sixth grade class in each school is taught using method 2.

Table 10.4. Means on a History of Ancient Civilizations Test Obtained by Sixth Grade Classes Using Two Different Instructional Methods

	Method 1	Method 2
Group A	20	15
Group B	25	35

An inspection of the four means listed in table 10.4 indicates that students in the first school (group A) who were taught using method 1 scored higher on the ancient civilizations test compared with their peers in the school who were taught using method 2. In the other school (group B), the results were reversed. In that school, students who were taught using method 2 scored higher than their peers who were taught using method 1.

The results reported in table 10.4 are displayed in figure 10.7. Although the two lines representing the two classes do not cross, they are on a "collision course," which is typical of a significant interaction effect.

When, in addition to having a significant interaction, the two main effects (the row variable and the column variable) are also significant, it may be difficult to interpret the results. Figure 10.8 illustrates another hypothetical example of interaction, showing the means of two groups (group A and group B) and two teaching methods (method 1 and method 2). As you can see, the interaction is significant (the lines cross). Group A scored higher than group B when method 2 was used, and group B scored a bit higher than group A when method 1 was used.

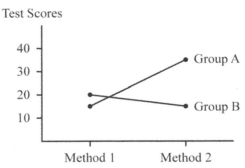

FIGURE 10.8
A graph showing a significant interaction and significant main effects

Looking at figure 10.8, we can speculate that in addition to a significant F ratio for the interaction, the two F tests for main effects (methods and groups) are also significant. As is often the case, those interpreting the results from the study should exercise caution when they make a decision about the efficacy of the teaching methods.

The Two-Way ANOVA Summary Table

As was mentioned, the results of the computations of a two-way ANOVA test are presented in a summary table, similar to the table that is used for presenting the results of a one-way ANOVA (see table 10.1). Each of the two factors in a two-way ANOVA (which are also called the *main effects*) is associated with an F ratio. Similarly, the interaction is analyzed using its own F ratio. As in the one-way ANOVA, a summary table is used to display the results of the two-way ANOVA analyses (see table 10.5). Note, though, that there are three F ratios in table 10.5, as opposed to a single F ratio in the one-way ANOVA (see table 10.1). The three F ratios in a two-way ANOVA are designed to test the three null hypotheses. Table 10.5 also shows the computations of the degrees of freedom associated with each of the three F tests. (The "Within Groups" listed in the *source* column may also be referred to as *Residual* or *Error*.)

Table 10.5. Two-Way ANOVA Summary Table

Source	SS	df	MS	F	p
Main Effects					
Factor 1 (row)	SS_R	no. of levels – 1	MS_R	F_R	<.05>
Factor 2 (column)	SS_C	no. of levels – 1	MS_C	F_C	<.05>
Interaction	SS_{RxC}	df_{row} x $df_{col.}$	MS_{RxC}	F_{RxC}	<.05>
Within Groups	SS_W	$N - K$	MS_W		
Total	SS_T	$N - 1$			

An Example of a Two-Way ANOVA

Two fourth grade teachers in Lincoln School want to know if the gender of the main character in a story makes a difference in their students' interest in stories they read. Each of the two fourth grade classes includes fourteen girls and fourteen boys; they are also fairly similar to each other in other student makeup, such as reading test scores.[6] The teachers choose two stories where the main character is a girl and two stories where the main character is a boy. The four stories are comparable in terms of plot, length, and readability level. Fourteen girls in one classroom and fourteen boys in the other classroom are asked to choose one of the two stories with a *girl* as the main character. The other fourteen boys and fourteen girls are asked to choose one of the stories with a *boy* as the main character. All students are asked to read their stories and complete an interest inventory designed to measure their interest in the stories read. Scores on the interest inventory can range from 10 to 50 points. Table 10.6 presents the scores of all fifty-six students, as well as the groups' means. The rows show the scores of the boys and girls in both classes, and the columns show the scores of the two groups of students who read the two types of stories (a story with a female main character and a story with a male main character).[7]

Table 10.6. Scores and Means of Boys and Girls on an Interest Inventory Following Reading Stories with Male vs. Female Main Characters

	Type of Story										
	Female as Main Character					Male as Main Character					Total
Boys	14	13	17	10	18	19	17	18	20	18	$\bar{X}_B = 16.46$
	16	15	15	15	16	17	16	18	20	19	
	14	17	16	15		18	17	16	17		
	$\bar{X} = 15.07$					$\bar{X} = 17.86$					
Girls	17	20	19	19	20	18	14	16	15	14	$\bar{X}_G = 16.39$
	18	16	15	17	19	13	17	15	14	14	
	18	16	17	18		13	16	15	16		
	$\bar{X} = 17.79$					$\bar{X} = 15.00$					
Total	$\bar{X}_F = 16.43$					$\bar{X}_M = 16.43$					

---

6. HINT: Although ANOVA assumes that the groups are random samples from their respective populations, in studies conducted in a typical school this assumption may not be fully met. As mentioned before, empirical studies have shown that ANOVA can be conducted under such conditions without seriously affecting the results, *especially when the group sizes are similar.*

7. HINT: Remember that in ANOVA, each mean represents a separate and independent group of individuals and no person can appear in more than one group. In our example, each student read *either* a story with a female main character *or* a story with a male main character (but *not both types* of stories).

Note that the total mean of the twenty-eight boys on the interest inventory ($\bar{x}_B$ = 16.46) is almost identical to the total mean of the twenty-eight girls ($\bar{x}_G$ = 16.39). There is also no difference in the students' interest in the two stories read (see the two column totals); the mean score for the boys and girls who read a story with a female character ($\bar{x}_F$ = 16.43) is identical to the mean score for the group of boys and girls who read a story with a male character ($\bar{x}_M$ = 16.43).

The interaction of student gender (the row factor) and the type of story read (the column factor) is displayed in figure 10.9. The figure shows that the lines cross, indicating a significant interaction of the two factors. There is a clear preference by boys to read stories where the main character is a male, while girls prefer stories with a female main character.

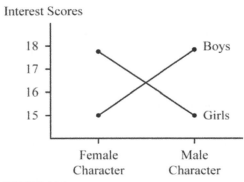

FIGURE 10.9
A graph showing the mean scores and significant interaction of the two factors (student gender and the gender of the main character) of the data in table 10.6

The two-way ANOVA summary table (table 10.7) lists the sum of squares (SS), the degrees of freedom (df), the mean squares (MS), the F ratios, and the level of significance (p value).[8]

Table 10.7. Two-Way ANOVA Summary Table

Source	SS	df	MS	F
Main Effects				
Gender	0.07	1	0.07	0.03
Type of Character	0.00	1	0.00	0.00
Interaction	108.64	1	108.64	43.13*
Within	131.00	52	2.52	
Total	239.71	55	4.36	

*$p < .001$

8. HINT: The "Within" (that is, within groups) listed under the "Source" column is also called *Residual* or *Error*.

The F ratios in the summary table confirm our earlier inspection of table 10.6. The F ratio for rows (comparing the interests of boys and girls) and the F ratio for columns (comparing the interest in the two types of stories) are not statistically significant ($F = 0.03$ and $F = 0.00$, respectively). However, the F ratio for the interaction effect is very high ($F = 43.13$) with a p value of $< .001$. We can be very confident about our decision to reject the null hypothesis. There is a strong interaction between the students' gender and their interest in the story they read.

If we had conducted two one-way ANOVA tests instead of the single two-way ANOVA test, we might have concluded the following: (a) there is no difference in the level of interest expressed by boys and girls toward the two types of stories (a nonsignificant F ratio for rows), and (b) the gender of the main character in the story does *not* make a difference in student's attitudes toward the stories (a nonsignificant F ratio for the columns). Upon inspecting the means in table 10.6, the interaction diagram in figure 10.9, and the level of statistical significance (p values) in table 10.7, it becomes clear that there *is* a difference between boys and girls in their level of interest toward the two types of stories. In our fictitious study, the boys were more interested in a story with a male as the main character, and the girls were more interested in a story with a female as the main character. If this were a real study, instead of a fictitious one, the implications for educators would have been that the selection of stories for their students should be made carefully, taking into consideration the gender of the main character in the story.

SUMMARY

1. An **analysis of variance (ANOVA)** is used to compare the means of two or more independent samples and to test whether the differences between the means are statistically significant.
2. The **one-way analysis of variance (one-way ANOVA)** can be thought of as an extension of a *t* test for independent samples. It is used when there are two or more independent groups.
3. The *independent* variable is the categorical variable that defines the groups that are compared. The *dependent* variable is the measured variable whose means are being compared.
4. There are several assumptions for ANOVA: (a) the groups are independent of each other, (b) the dependent variable is measured on an interval or ratio scale, (c) the dependent variable being measured is normally distributed in the population, (d) the scores are random samples from their respective populations, and (e) the variances of the populations from which the samples were drawn are equal (the assumption of the *homogeneity of variances*). The first two assumptions cannot be violated.
5. ANOVA is considered a *robust* statistic that can stand some violation of the third and fourth assumptions. Empirical studies show that there are no serious

negative consequences if these assumptions are not fully met. The assumption of the homogeneity of variances can be tested using special tests.

6. The test statistic in ANOVA is called the F statistic. The **F ratio** (or **F value**) is computed by dividing two variance estimates by each other. If the F ratio is statistically significant (that is, if $p < .05$) and if there are three or more groups in the study, then a pairwise **post hoc comparison** is done to assess which means are significantly different from each other.

7. When one independent variable is used, the design is called a *one-way analysis of variance*; when two independent variables are used, the design is called a **two-way analysis of variance**. In general, when two or more independent variables are used, the design is called a **factorial ANOVA**.

8. In the factorial ANOVA test, the **interaction** refers to a situation where one or more levels of the independent variable have a different effect on the dependent variable when combined with another independent variable.

9. An independent variable must have at least two *levels* (or conditions).

10. The *null hypothesis* (H_0) in a one-way ANOVA states that there is no significant difference between the population means; the *alternative hypothesis* (H_A) states that at least two population means differ significantly from each other.

11. The *variability* in a one-way ANOVA is divided into three **sums of squares** (SS): *within-groups* sum of squares (SS_W), *between-groups* sum of squares (SS_B), and *total* sum of squares (SS_T). The SS_W is equal to the combined SS and SS_B.

12. The variance estimates in ANOVA are called the **mean squares**. The mean squares *between* (MS_B) and the mean squares *within* (MS_W) are obtained by dividing SS_B and SS_W by their corresponding degrees of freedom.

13. The MS_W (also called the *error term*) can be thought of as the average variance to be expected in any normally distributed group; it serves as the denominator in the computation of the F ratio.

14. The F ratio is obtained by dividing the MS_B by the MS_W:

$$F = \frac{MS_B}{MS_W}$$

15. The results of the ANOVA test are often displayed in a *summary table*. The summary table includes the sum of squares, degrees of freedom, mean squares, F ratio, and level of significance (p value).

16. If the F ratio is statistically significant ($p < .05$), a *post hoc comparison test*, such as *Tukey's honestly significantly difference (HSD)* test, is conducted to determine which means are significantly different from each other.

17. The **two-way ANOVA** test is used to compare two independent variables (or factors) *simultaneously*. This statistical test enables us to study the effect of each of the two factors on the dependent variable as well as the interaction of the two factors. The independent variables in factorial ANOVA are also called the main effects.

18. A two-way ANOVA is conducted to test three hypotheses about differences between the *row* variable, the *column* variable, and the *interaction* of these two independent variables (or factors). Three *F* ratios are calculated to test each of the three null hypotheses.

19. The *total* variation in a two-way ANOVA is partitioned into two main sources: the *within-groups* variation and the *between-groups* variation. The *between-groups* variation is further partitioned into three components: the variation among the *row mean*, the variation among the *column means*, and the variation due to *interaction*.

20. As was the case with a one-way ANOVA, the two-way ANOVA summary information is presented in a table. The summary table includes four sums of squares, four degrees of freedom, three mean squares, three *F* ratios, and three *p* levels.

21. In order to better understand an interaction, it is often helpful to graph it. To create the *interaction graph*, the mean scores on the dependent variable are marked on the vertical axis. Lines are then drawn to connect the means of the groups.

22. A *nonsignificant* interaction is represented by *parallel* lines, and a *significant* interaction is represented by *nonparallel* lines.

23. There are two types of significant interactions: (a) *disordinal*, where the lines intersect; and (b) *ordinal*, where the lines do not intersect (but are not parallel).

CHECK YOUR UNDERSTANDING

1. What are the similarities and differences between *t test for independent* samples and *one-way ANOVA*? List at least two differences and two similarities.

2. Design a study where one-way ANOVA may be used with a minimum of three groups. Describe your study and explain your choice of one-way ANOVA to analyze the data.

3. How many *F ratios* are calculated for *two-way ANOVA* and why?

4. In what types of ANOVA studies would it be helpful to study the *interaction* of the *independent* variables? Explain.

5. Design a study where *two-way ANOVA* may be used to analyze the data. Describe your study and explain your choice of this statistical test (that is, two-way ANOVA).

V

CHI SQUARE TEST

Chi Square Test

Chapter 11 introduces you to the *chi square test*, represented by the Greek letter *chi*, squared (χ^2). This test uses frequencies ("head counts") from discrete, categorical data, and it is the only one in the book that belongs to the type of statistics that are called *nonparametric*. (Nonparametric statistics can be applied to data that do not meet other, more strict requirements that are necessary for *parametric* statistics. Correlation and *t* test, for example, are considered parametric statistics.) The frequencies that are counted are called the *observed* frequencies, and they are compared to *expected* frequencies. The chi square test statistic is then computed and evaluated to determine whether it is statistically significant.

This chapter presents two kinds of chi square tests. The *test of best fit* is used when there is *one independent variable* (for example, comparing responses of girls to boys on a question that requires a Yes/No answer where gender is the independent variable). The second test is called the *test of independence*, and it is used when there are *two independent variables* (for example, comparing the responses of administrators, teachers, and parents from two neighboring school districts to a question that requires a Yes/No answer; role and district are the independent variables). Numerical examples in this chapter will help you gain an appreciation and understanding of the chi square test.

When the data collected do not meet the requirements necessary for parametric statistics, we can use statistical tests that belong to a group of statistical methods called *nonparametric*, or distribution-free. These statistical tests can be applied to data that do not meet certain assumptions (for example, being measured on an interval or ratio scale, or being normally distributed). (See chapter 2 for a discussion of measurement scales.) The **chi square** (χ^2) test is applied to discrete data (that is, nominal, categorical data). The units of measurement that are often used are frequency counts and observations (rather than scores). The chi square test was developed by Karl Pearson (who also developed the Pearson product moment correlation) in 1900 in order to measure how well observed data fit a theoretical distribution.

UNDERSTANDING THE CHI SQUARE TEST

The chi square statistic can be used to analyze data measured on a *nominal* scale, such as gender, where there are two or more discrete categories. It can also be used to analyze other types of numerical data (such as data measured on an interval scale) that are first divided into *logically defined* and generally agreed-upon categories. For example, IQ scores can be divided into three categories (high, average, and low) by using the standard deviation of the IQ scale to define each category.

The chi square test is often used to analyze questionnaire data where a numerical code is assigned to groups or responses. For example, the political affiliations of registered voters may be assigned a numerical code of 1–4 as follows: *Democrats* = 1; *Republicans* = 2; *Independents* = 3; and *Other* = 4.[1] In another example, a number is assigned to each position in the school: teacher is coded as 1; administrator is coded as 2, and support staff is coded as 3.

In applying the chi square test, two types of frequencies are used: *observed* and *expected*. The *observed frequencies* (O) are based on actual (empirical) observations and on "head counts." An example of observed frequencies is the actual number of people who respond "Yes" or "No" to a particular question. The *expected frequencies* (E) are theoretical or based on prior knowledge. The observed and expected frequencies can be expressed as actual head counts or as percentages. (The process for determining the expected frequencies is explained later in this chapter.) The chi square test is used to decide whether there is a significant difference between the observed and expected frequencies, and both types of frequencies are used in the computation of the *chi square value* (χ^2).

Each pair of observed frequencies and its corresponding expected frequencies is called a *cell*. To compute χ^2, for each cell we start by computing $(O - E)^2$, then divide it by E. We then add up the results of the computations from each cell to obtain the χ^2 value. The formula for χ^2 is:

$$x^2 = \sum \left[\frac{(O-E)^2}{E} \right]$$

1. HINT: Political affiliation is a nominal scale variable, and the code we assigned in this example was arbitrarily chosen.

2. HINT: When one of the variables is *group membership* (for example, gender or political affiliation), and the other variable is the responses of the *group members* (for example, "yes" or "no"), the convention is to record the groups in the *rows* and their responses in the *columns*.

Where χ^2 = Chi square statistic
    O = Observed frequencies for each cell
    E = Expected frequencies for each cell

The numerator in the equation includes the difference between the *observed* and *expected* frequencies. When there are very small differences between the observed and expected frequencies in each cell, the numerator is small and the chi square value is low. When there are large differences between the observed and expected frequencies in each cell, the numerator is large, and, in turn, so is the chi square value.

The degrees of freedom (*df*) in the chi square statistic are related to the number of *levels* (that is, categories) in the dependent variable(s). This is different from the procedures for computing the degrees of freedom in other statistical tests, such as Pearson correlation and *t* test, where the degrees of freedom are related to the sample sizes. This chapter includes explanations of the proper procedures for calculating the degrees of freedom used in the chi square analysis.

ASSUMPTIONS FOR THE CHI SQUARE TEST

Various types of data can be analyzed using the chi square statistic. Several assumptions are required in order to apply the chi square test. These assumptions are:

1. The observations should be *independent* of each other, and a particular response cannot be counted in more than one category. For example, a person may not be asked on two different occasions to respond to the same question, as if two people each responded once. The total number of observed frequencies should not exceed the number of participants.
2. The data must be in the form of *frequencies* and the total number of observed frequencies must equal the total number of expected frequencies.
3. The categories, especially those that comprise ordinal, interval, or ratio scale, should be created in some *logical*, defensible way. The criteria used to establish the categories should be chosen carefully and wisely. For example, suppose one of the variables in a study is the participants' level of income. Income is a continuous variable, and it is necessary to establish logical cutoff points to define the various categories. The researcher may want to divide the variable of *income* into categories (for example, *high*, *middle*, and *low*) by following some official guidelines for defining income levels.

The chi square *goodness of fit test* is used to test the fit between a distribution of observed frequencies and a distribution of expected frequencies. The chi square *test of independence* is used to test whether two factors (independent variables) are independent of each other. In both types of chi square, observed frequencies are compared to expected frequencies.

THE CHI SQUARE GOODNESS OF FIT TEST

In a goodness of fit chi square test, the number of expected frequencies in each category may be *equal* or *unequal*. The following is a discussion of the two types.

Equal Expected Frequencies

In this type of chi square test, there are *equal expected frequencies* in each category. The observed frequencies, as always, are based on empirical data—that is, on observation. We collect data by recording the number of occurrences in each category. For example, we can use the chi square procedure to test whether a coin is fair by tossing the coin 100 times and recording the number of heads and tails. These numbers are our *observed* frequencies. The *expected* frequencies are based on the assumption that the coin is fair; thus, half of the time it should land heads, and half of the time, tails. The null hypothesis states that the coin is fair and, consequently, would land as many times heads as tails.

Suppose we toss a coin 100 times and record 55 heads and 45 tails (see table 11.1). With a fair coin (the null hypothesis is true), we would expect 50 heads and 50 tails. Therefore, we must ask ourselves whether our coin is fair or biased. As with other statistical tests, such as the *t* test, we cannot simply eyeball our observed data and decide whether the coin is biased based on our observations. In other words, it would be difficult to look at the difference between 55 and 45 and determine whether it is large enough to indicate a biased coin or whether this difference is small enough to have occurred purely by chance.

Table 11.1. Observed and Equal Expected Frequencies for Heads and Tails

	O (Observed)	E (Expected)	$\dfrac{(O - E)^2}{E}$
HEADS	55	50	$\dfrac{(55 - 50)^2}{50} = \dfrac{25}{50} = 0.50$
TAILS	45	50	$\dfrac{(45 - 50)^2}{50} = \dfrac{25}{50} = 0.50$
TOTAL	100	100	$\chi^2 = 1.00$

The computations that are displayed in table 11.1 show that we obtained a chi square value of 1.00. We now need to determine whether this test statistic is statistically significant by consulting the table of critical values for chi square statistic (see table 11.2). The degrees of freedom (*df*) are the number of categories, or levels, minus

1. In our example, the variable has two levels, heads and tails, resulting in a *df* of 1 (2 − 1 = 1).

Table 11.2. Partial Distribution of the χ^2 Critical Values (df = 1)

		p values		
df	.10	.05	.02	.01
1	2.706	3.841	5.412	6.635

As before, unless told otherwise, we use the convention of 95 percent confidence level and inspect the critical values listed in the column of *p* = .05. With 1 degree of freedom (*df* = 1), the appropriate critical value is 3.841. Our obtained χ^2 value of 1.00 *does not* exceed this critical value; as a result, we *retain* the null hypothesis. Our conclusion is that the coin is fair even though when we tossed it 100 times it landed more times heads than tails. The difference between the numbers of heads and the tails is small enough to have happened by chance alone and is probably due to a random error, rather than a systematic error (that is, a biased coin).

Unequal Expected Frequencies

The chi square test for *unequal expected frequencies* is used primarily to examine similarities and differences between a group's *observed* frequencies and its *expected* frequencies that are *unequal*. Often, the expected frequencies represent the population distribution of the observations on the variable being investigated. In the study, the researcher studies the match between the sample distribution (the observed frequencies) and the population distribution (the expected frequencies). The researcher has to know the expected frequencies *a priori* (ahead of time) in order to conduct this type of chi square analysis. An example may help to illustrate this chi square test.

A series of articles published in the local press reports that there is an ongoing problem of grade inflation in the School of Education at the state university. Reporters contend that too many grades of A and B are given to undeserving students in the School of Education. The dean of the school argues that the distribution of grades given to students in the school is comparable to the distribution of grades in other similar institutions. The chi square statistic is selected to analyze the data and compare the distribution of grades in the School of Education (the observed frequencies) with the grades in other similar institutions (the expected frequencies) (table 11.3). The null hypothesis states that there is no difference in the distribution of grades between the School of Education and other similar institutions. If the two distributions of observed and expected frequencies turn out to be similar, the

Table 11.3. Observed and Unequal Expected Frequencies for Five Letter Grades

Grade	O	E	$\frac{(O - E)^2}{E}$
A	16	10	$\frac{(16 - 10)^2}{10} = 3.60$
B	22	20	$\frac{(22 - 10)^2}{20} = 0.20$
C	38	40	$\frac{(38 - 40)^2}{40} = 0.10$
D	17	20	$\frac{(17 - 20)^2}{10} = 0.45$
F	7	10	$\frac{(7 - 10)^2}{10} = 0.90$
TOTAL	100	100	$\chi^2 = 5.25$

resulting chi square value would be small, thus leading the researchers to retain the null hypothesis. As was discussed before, in order to be able to conduct this type of chi square test, the researcher has to have *a priori* knowledge about the distribution of the expected frequencies.

According to table 11.3, the obtained chi square value is 5.25. The degrees of freedom in this example are 4 ($df = 5 - 1 = 4$). Table 11.4 shows the critical values for χ^2 values with 4 degrees of freedom.

Table 11.4. Partial Distribution of the χ^2 Critical Values ($df = 1$)

		p values		
df	.10	.05	.02	.01
4	7.779	9.488	11.668	13.277

The critical value of χ^2 at $p = .05$ and df of 4 is 9.488. Our obtained value of 5.25 *does not* exceed this critical value, leading us to *retain* the null hypothesis. Further inspection of the data shows that in comparison with other similar institutions, more grades of A and B and fewer grades of D and F were given in the School of Education. However, these differences are not statistically significant and do not indicate a great departure from the "norm." The school's dean may still want to review the grading process in the school, as they seem to differ somewhat from the standards at other colleges.

THE CHI SQUARE TEST OF INDEPENDENCE

The chi square *test of independence* is conducted to test whether two independent variables (or factors) are related to, or are independent of, each other. For example, a researcher may want to investigate whether there is a difference in the political party affiliation between teachers and parents in the school district. The researcher may survey 100 teachers and 100 parents, asking them to indicate whether they are Democrat, Republican, or Independent.[2] The responses of all participants are then tallied and arranged in a 2 × 3 ("two by three") table (see table 11.5).

Table 11.5. Observed Frequencies: Teacher and Parent Survey About Political Party Affiliation

	Democrats	Republicans	Independents	TOTAL
Teachers	60	35	5	100
Parents	50	45	5	100
TOTAL	110	80	10	200

The null hypothesis states that political party affiliation is independent of group membership (that is, there is no difference in the political affiliation distribution between the teachers and the parents). The degrees of freedom for the chi square test of independence are calculated as the number of rows minus 1, multiplied by the number of columns minus 1: (row − 1)(column − 1). In the example above, there are two rows and three columns, so the degrees of freedom are 2 ($df = [2 − 1] \times [3 − 1] = 2$).

The most common tables in the chi square tests are those that have two levels in each of the two variables (for example, boys/girls and yes/no). These are *contingency tables*, which are referred to as *2 × 2* ("two by two") *tables*. Next, we use an example to take you through several of the steps in the computations of a chi square value using data presented in a 2 × 2 table. In this example, a group of eighty regular students and ninety English Language Learners (ELL) students are asked to respond to the following question: "I believe my classmates like me" by circling "yes" or "no." Table 11.6 displays their responses (the observed frequencies). The observed and expected frequencies are paired in cells, referred to as A, B, C, and D.

Table 11.6. Observed Frequencies: Regular and English Language Learners and Their Responses to the Question: "I Believe My Classmates Like Me"

	Yes	No	TOTAL
Regular Students	(Cell A) 62	(Cell B) 18	80
ELLs	(Cell C) 57	(Cell D) 33	90
TOTAL	119	51	170
			⇑
			Grand Total

Even without conducting any statistical test, it is clear that the majority of students in both groups perceive that their classmates like them. Nevertheless, just by inspecting the data in the table it is difficult to know whether any differences we observe between the groups are indicative of significantly different self-perceptions or are due to some sampling error and could have happened purely by chance. As before, we cannot simply eyeball the data and come up with a conclusion.

The *expected* frequencies are based on the null hypothesis that states that there is no relationship between the group membership ("regular" versus "ELL") and the students' perceptions. In other words, the null hypothesis states that the two variables (group membership and perception) are independent of each other. Table 11.7 lists the observed and expected frequencies and the totals. The computations of the expected frequencies are a bit tedious and are not included here. Computer statistical packages, such as SPSS, can easily provide the expected frequencies as part of the chi square analysis in their cross-tabulation tables.

Table 11.7. Observed and Expected Frequencies: Regular and English Language Learners and Their Responses to the Question: "I Believe My Classmates Like Me"

	Yes		No		
Group	Observed	Expected	Observed	Expected	TOTAL
Regular Students	62	56	18	24	80
ELLs	57	63	33	27	80

The observed row and column totals are used to compute the expected frequencies for each cell. Note that in table 11.7, the row and column *totals* for the expected frequencies are the same as those for the observed frequencies. As before, we use the observed and expected values in each cell to compute a cell value. The values from each cell are then added to obtain a chi square value of 4.04 (0.64 + 1.50 + 0.57 + 1.33 = 4.04).

Table 11.8. Computation of the Cell Values

$$\frac{(O-E)^2}{E}$$

$$\text{Cell A}: \frac{(62-56)^2}{56} = 0.64 \qquad \text{Cell B}: \frac{(18-24)^2}{24} = 1.50$$

$$\text{Cell C}: \frac{(57-63)^2}{63} = 0.57 \qquad \text{Cell D}: \frac{(33-27)^2}{27} = 1.33$$

The degrees of freedom are calculated as $df = (2 - 1)(2 - 1) = 1$. Next, we consult table 11.8 that shows the critical values associated with one degree of freedom.

Table 11.9. Partial Distribution of the c^2 Critical Values ($df = 1$)

	p values			
df	.10	.05	.02	.01
1	2.706	3.841	5.412	6.635

Our obtained χ^2 value of 4.04 *exceeds* the critical value under p of .05 and 1 *df,* which is 3.84, but not the value under $p = .02$, which is 5.41. Therefore, we report the results of this chi square test to be significant at $p < .05$ and *reject* the null hypothesis. We are at least 95 percent confident that our decision to reject the null hypothesis is the right decision. The differences in the responses of the two groups of students are too large to have occurred purely by chance. We conclude that although in both groups the majority of students feel their classmates like them, the percentage of regular education students responding positively to the question posed is higher than that of the ELL students. The results of our study suggest that students' responses may depend on their group membership.

SUMMARY

1. The **chi square (χ^2) test** is applied to discrete, categorical data where the units of measurement are frequency counts.
2. The chi square test is considered a *nonparametric*, or a *distribution-free* statistic. It can be used to analyze data measured on a *nominal* scale (such as gender) where there are two or more discrete categories. It can also be used to analyze other types of numerical data (such as data measured on an interval scale) that are first divided into *logically defined* and generally agreed-upon categories.
3. The chi square test is often used to analyze questionnaire data where a numerical code is assigned to groups or types of responses.
4. In applying the chi square test, two types of frequencies are used: *observed* and *expected*. The observed frequencies are based on actual (empirical) observations and on "head counts." The expected frequencies are theoretical or based on prior knowledge. The observed and expected frequencies can be expressed as actual head counts or as percentages.
5. Each pair of observed frequencies and its corresponding expected frequencies is called a *cell.*
6. The observed and expected frequencies in each cell are used to compute the chi square value. In the formula for computing the chi square, "O" refers to the observed frequencies and "E" refers to the expected frequencies in each cell. The chi square value is found by adding up the results of the division (the quotient) from each cell.

$$\chi^2 = \sum \frac{(O-E)^2}{E}$$

7. The *degrees of freedom* (*df*) in the chi square statistic are related to the number of *levels* or *cells* (that is, categories) in the independent variable(s). This is different from the procedures for computing the degrees of freedom in several other statistical tests, such as Pearson correlation and *t* test, where the degrees of freedom are related to the sample sizes.

8. The assumptions required for applying the chi square statistic are: (a) the observations should be *independent* of each other, (b) the data are recorded as *frequencies* ("head count"), and (c) the categories are created in some *logical* and agreed-upon way.

9. There are two types of chi square tests: (a) The *goodness of fit test*, with one independent variable, is used to test the fit between a distribution of observed frequencies and expected frequencies; and (b) the *test of independence*, with two independent variables, is used to test whether the two factors (the independent variables) are independent of each other.

10. In a chi square *goodness of fit* test, the number of expected frequencies in each category may be *equal* or *unequal*. When *equal* expected frequencies are used, they represent the null hypothesis that posits that there is an equal probability of having the same number of frequencies in each cell. When *unequal* expected frequencies are used, they must be known *a priori* (ahead of time) and be provided by the researcher.

11. The chi square *test of independence* is used to determine whether two variables are related to, or are independent of, each other. Each of the two variables has to have at least two levels (for example, male/female, true/false, above/ below).

CHECK YOUR UNDERSTANDING

1. Explain at least two differences between *chi square* and *t test*, or between *chi square* and *ANOVA*.

2. Explain the differences between chi square *test of best fit* and chi square *test of independence*.

3. Propose two studies where chi square *goodness of fit* test or chi square *test of independence* are used to analyze the data, address the following:

    a. Illustrate and describe the study, including your choice of chi square to analyze the data.

    b. Create a grid (for example, 2x2) to illustrate the data.

    c. Explain how you will determine the *statistical significance* of the data and the type of conclusions that you may be able to draw.

4. When a chi square test study represented in a 3x5 table is reported to be *statistically significant*, what exactly does it mean? Explain.

VI

STANDARDIZED TEST SCORES, RELIABILITY, AND VALIDITY

Interpreting Standardized Test Scores

Being in a school system, you are probably aware of the major role of testing. Chapter 12 focuses on two main types of tests that are most prevalent in schools: norm-referenced and criterion-referenced tests. Many of the high-stakes tests that are administered to schoolchildren, such as mandated state tests, are norm-referenced. Other norm-referenced tests are admission tests used by universities (for example, SAT, ACT, and GRE). The word *norm* refers to the norming group that was used in developing the test. The performance of test-takers is compared to those in the norming group through the use of scale scores, percentiles, stanines, and grade equivalents. Norm-referenced tests are usually created by professional test writers.

By comparison, criterion-referenced tests are used to compare the performance of students to certain criteria, not to each other. Criterion-referenced tests are available commercially and may also be constructed by classroom teachers.

Tests are used in all areas of life. They are given to people seeking certifications and, in some cases, to job applicants. They are also used to determine placement and admission into programs, to diagnose and evaluate patients, to monitor progress, to assign grades, and more. In this chapter, we focus our attention on standardized tests that are used extensively in education: norm-referenced tests and criterion-referenced tests.[1]

STANDARDIZED TESTS IN EDUCATION

Those who create educational tests include classroom teachers, state boards of education, textbook writers, and national corporations. Some of the biggest designers of tests are commercial companies that produce school-related standardized achievement tests, such as the Terra Nova (published by CTB McGraw-Hill). The enactment of the No Child Left Behind bill, passed by legislators in Washington in 2001, placed a heightened emphasis on mandatory testing of school-age children. Starting in 2013, a formal large-scale assessment of the Common Core State Standards was implemented across most states.

Although there are a variety of tests and assessment tools administered to school-age children, our discussion in this chapter focuses exclusively on school-related tests, particularly achievement tests.

There are several ways to report test scores. Some of the most common ways are *raw scores, percent correct, percentile ranks, stanines, grade equivalents*, and *scale scores*. (See chapter 6 for a discussion of percentile ranks.)

It is hard to interpret raw scores obtained by students on an achievement test if no additional information is available about the test, such as the number of items and their level of difficulty, or the scores of the other examinees who took the tests. Classroom teachers often convert the raw scores obtained by their students on a teacher-made test to percent correct and letter grades. Raw scores derived from standardized tests constructed by commercial test companies are usually converted into scale scores and norms.

Standardized tests can be classified into two major categories: *norm-referenced* and *criterion-referenced*. The two types of tests differ in the way they are constructed and how they are used.

NORM-REFERENCED TESTS

Norm-referenced (NR) tests include norms that allow the test user to compare the performance of an individual taking the test to that of similar examinees who have taken the test previously. These examinees comprise the *norming group*. The **norming group** is a sample taken from the population of all potential examinees. A stratified random sampling procedure is usually used to select the sample for norming. Stratification is done on characteristics such as gender, age, socioeconomic status, race, and geographic region. The norming group should be large enough and demographically represent the characteristics of the potential test-takers. The test is first given to the

---

1. HINT: A discussion of teacher-made tests and other forms of assessment is outside the scope of this book.

norming group, and then the scores on the test are used to generate the norms. Later, when new examinees take the test, their scores are usually compared to the scores of the norming group, rather than to the scores of others taking the test with them. However, in some cases the score of an examinee is compared to the scores of those who took the test at the same time in order to generate local norms.

In standardized tests, items are first pilot tested and revised, as necessary. Test items constructed for NR tests are written specifically to maximize differences among the examinees. Some items have a high level of difficulty in order to differentiate among the top students, while other items are easy in order to distinguish among the low-scoring students. Easy items may also be placed at the beginning of the test or section to encourage all students. Most items are of average difficulty and are designed to be answered correctly by 30 to 80 percent of the examinees.

Commercial achievement test companies describe in their technical manual how the norming group was selected, its demographic characteristics, and when the norms were obtained. Other technical aspects of the test, such as its reliability and validity, are likely to be discussed in the manual as well. Norms are usually provided for standardized, commercially constructed tests.

Testing companies may develop two types of norms: *national* and *local*. This is especially common with standardized achievement tests, which are administered annually to students across the United States. In a typical school, students and their parents receive a computer-generated report that lists the raw scores as well as national and local norms. The *national norms* compare each student to similar students in the population at large, while *local norms* compare the student to others with the same demographic characteristics, such as other students in the district or school.

Several tests, such as college admission tests, are designed for a particular purpose. The norming group, although more specific, is still comprised of examinees with characteristics similar to those of the potential test users. For example, the Scholastic Aptitude Test (SAT) and the ACT assessment are normed on college-bound high school juniors or seniors. The Graduate Record Examination (GRE) is normed on students who plan to attend graduate schools. Professional graduate programs (such as law schools, business schools, and medical schools) have their own admission tests that are normed on a representative sample of students who apply to these professional programs.

Test publishers report several kinds of scores. A typical student report includes raw scores on each subtest, in addition to norms. Three of the most commonly used norms are *percentile ranks*, *stanines*, and *grade equivalents (GE)*.

Percentile Ranks

A *percentile rank* describes the percentage of people who scored *at* or *below* a given raw score. For example, when a raw score of 58 is converted to a percentile rank of 82, it means that a student with that raw score performed better than, or as well as, 82 percent of those in the norming group. At times, a percentile rank is described simply as the percentage of examinees that scored *below* a given score (omitting the word *at* from the definition). Standardized achievement test publishers routinely include percentile ranks in the reports they provide to school personnel and parents. (See chapter 6 for a more comprehensive discussion of percentile ranks and percentiles.)

Percentile ranks are easy for laypeople to understand, which may be one of the reasons they are popular as norm-referenced scores. The public may also be familiar with the concept of percentile ranks because they are used by pediatricians to chart the height and weight of babies and young children.

In addition to percentile ranks, standardized test reports often include percentile bands. Since the tests are not completely reliable and include a certain level of error, the band gives an estimated range of the true percentile rank. A confidence level of 68 percent is commonly used in constructing the band. On the test report, the band is often represented by a shaded area.

After a commercial norm-referenced achievement test is administered in school, the parents or guardians of the students are likely to receive reports describing their children's performance on the test. Although the format and content of the reports produced by various testing companies differ from each other, most of them include information about the student's national percentile ranks and percentile bands on the subject areas covered by the test.

Additional information provided on test reports may include the following: local percentile ranks, a breakdown of the various subject areas into subscales, the total number of items for each subscale, and the number of items answered correctly by the student. To help parents understand the report, an explanation of the information is usually provided. In addition, parents are encouraged to meet with their children's teachers, who can provide further explanation of the report.

Stanines

The word **stanine** was derived from the words "*stan*dard *nine*." Stanines comprise a scale with 9 points, a mean of 5, and a standard deviation of 2. In a bell-shaped distribution, stanines allow the conversion of percentile ranks into nine larger units (see figure 12.1). Thus, stanine 5 includes the middle 20 percent of the distribution; stanines 4 and 6 each include 17 percent; stanines 3 and 7 each include 12 percent; stanines 2 and 8 each include 7 percent; and stanines 1 and 9 each include 4 percent of the distribution. Approximately one-fourth of the scores in the distribution (23 percent, to be exact) are in stanines 1–3, and 23 percent of the scores are in stanines 7–9. Approximately one-half (54 percent) of the scores in the distribution are contained in stanines 4–6, the middle stanines.

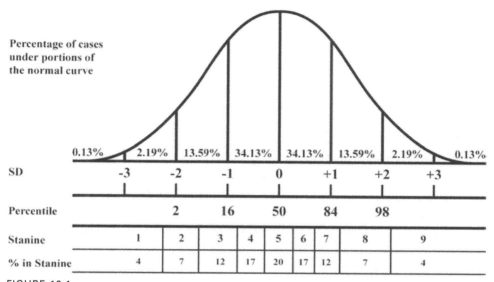

FIGURE 12.1

A normal distribution graph showing standard deviations, percentiles, stanines, and percentages of scores in each stanine

Grade Equivalents

Grade equivalents (GE) are used to convert raw scores into grade-level norms, expressed in terms of *years* and *months*. GE consists of a *whole number* representing the grade level, and a *tenth*, representing the month of the school year. For example, a GE of 4.2 is equivalent to the average raw score obtained by students in the norming group at the end of the second month of the fourth grade. "Typical" students are expected to gain 1 GE a year to maintain their position in relation to their age-mates.

Grade equivalents from different tests cannot be accurately compared to each other because each GE is derived from a different test. As a result of the way they are derived and calculated, grade equivalents should not be averaged for groups.

Grade equivalents are often misunderstood and misinterpreted. Therefore, there are schools that do not include the student's GE scores in the report to parents. As an example, parents of a fifth grader who receives a GE of 7.5 in reading may request that their child be promoted to the seventh grade. This request is misguided for several reasons: (a) the other students in the fifth grade class may also be performing above grade level; (b) the student may be above grade level in reading, but not in other subjects; (c) the fifth grader may have difficulties adjusting socially to peers in the seventh grade; and (d) a GE of 7.5 means only that the average student in the seventh grade would have received that score, not that a fifth grader is likely to succeed in the seventh grade.

In assigning a GE of 7.5 to a particular raw score on the fifth grade test, those who developed the norms are actually speculating on how well a seventh grader would have performed on the fifth grade test because the seventh graders in the norming group most likely did not actually take the fifth grade test. In addition, since the fifth grader has not actually taken the seventh grade test, we should not assume that this student is likely to succeed in the seventh grade.

CRITERION-REFERENCED TESTS

Criterion-referenced (CR) tests are designed to compare the performance of an individual to certain criteria (unlike norm-referenced tests that compare the examinees to other people in the norming group).[2] The criteria, which should be specific and clear, are based on skills or objectives as set forth by educators (for example, teachers, curriculum specialists, and content experts). After specifying the criteria, a task is designed to measure the extent to which the criteria have been met. For example, the task may be a pencil-and-paper or computer-assisted achievement test, or it may be performance-based, such as identifying countries on a world map.

Two main types of scores are used with CR tests: *percent correct* and *mastery/nonmastery*. Reporting scores in terms of percent correct is often used by the classroom teachers to generate letter grades. For example, a teacher may inform the class that to get an A on a test, one must score at least 90 percent correct, and to get a B, one must score 80 to 89 percent correct. This type of score does not take into consideration that the whole test may be too difficult or too easy.

Reporting scores in terms of mastery/nonmastery is based on the theory of mastery learning that advocates mastery of the present material before moving on to new material. There are several approaches that can be used to set the standards for mastery and to determine the point separating mastery from nonmastery. For example, content specialists can help establish a cutoff score to separate mastery from nonmastery, while other educators may decide that students have to answer at least 80 percent of the items correctly in order to demonstrate mastery.

While several published CR tests are available in specific areas, such as mathematics or reading, many publishers also include CR interpretation in their NR tests. In addition to listing information about norms, such as percentiles and stanines, the computer-generated report sent to the school for each student may also show the total number of items in each section of each subtest and the number of items answered correctly by the student. Similar information about the whole class may also be included. This information can help the teacher diagnose the strengths and weaknesses of individual students as well as the whole class.

Another way used by states to report on students' performance is to classify their test scores as falling into one of four categories: "academic warning," "below expectation," "meet expectation," and "exceed expectations."

---

2. HINT: Criterion-referenced tests may also be called *domain-referenced* or *content-referenced* tests.

SUMMARY

1. There are several ways to report test scores. Some of the most common ways are: *raw scores, percent correct, standard scores* (such as *z scores*), *percentile ranks, stanines, grade equivalents,* and *scale scores.*

2. Raw scores derived from teacher-made tests are usually converted into letter grades; raw scores derived from standardized tests are usually converted into norms.

3. Tests can be classified into two major categories: *norm-referenced* and *criterion-referenced.* These two types differ in the way they are constructed and how they are used.

4. **Norm-referenced (NR) tests** include norms that allow the test user to compare the performance of an individual taking the test to that of similar examinees who have taken the test previously.

5. A **norming group** is a sufficiently large sample with demographic characteristics similar to those of potential test takers. Scores from the norming group are used to develop the test norms. When new examinees take the test, their scores are compared to the scores of the norming group.

6. Test items constructed for NR tests are written specifically to maximize differences among the examinees. Some items have a high level of difficulty in order to differentiate among the top students, while other items are easy in order to distinguish among the low-scoring students. Most items are of average level of difficulty and are designed to be answered correctly by 30 to 80 percent of the examinees.

7. *Technical manuals* of standardized tests include information about the test development process, the demographic characteristics of the norming sample, and other psychometric information (such as the test reliability and validity).

8. Testing companies may develop two types of norms: *national* and *local.* National norms compare the student to similar students in the population at large, while local norms compare the student to others with the same demographic characteristics, such as other students in the district or school.

9. A *percentile rank* describes the percentage of people who scored *at* or *below* a given raw score. Percentile ranks are easy for laypeople to understand, which may be one of the reasons they are popular as norm-referenced scores.

10. A **percentile band** is often used to provide an estimated range of the true percentile rank. The bands are used due to the fact that the tests are not completely reliable and include a certain level of error.

11. **Stanines** (derived from the words "*stan*dard *nine*") comprise a scale of norms that is used to convert percentile ranks into nine larger units. The scale has a mean of 5 and a standard deviation of 2.

12. **Grade equivalents (GE)** are used to convert raw scores into grade-level norms. A grade equivalent consists of a number that represents the grade level and the month of the school year.

13. Grade equivalents from different tests cannot be accurately compared to each other because each GE is derived from a different test. Grade equivalents are often misunderstood and misinterpreted.

14. **Criterion-referenced (CR) tests** are designed to compare the performance of an individual to certain criteria. The criteria, which should be specific and clear, are based on skills or objectives as set forth by educators. Two main types of scores are used with CR tests: *percent correct* and *mastery/nonmastery*.
15. Reporting scores in terms of percent correct is easy to do and to understand. It is often used by classroom teachers to generate letter grades.
16. Reporting scores in terms of mastery/nonmastery is based on the theory of mastery learning that advocates mastery of the present material before moving on to new material.

CHECK YOUR UNDERSTANDING

1. What are the differences between *norm-referenced* and *criterion-referenced* tests? List at least two differences.
2. What are the *advantages* and *disadvantages* of norm-referenced and criterion-referenced tests?
3. List one example of a norm-referenced test and one example of a criterion-referenced test that are used in your school or other educational settings.
   a. Describe these tests briefly, including their purpose and use.
   b. What types of scores are reported in each of these tests? How are the results communicated to teachers or parents?

Reliability

Chapter 13 discusses the concept of *reliability* as it applies to tests and measurements. Reliability refers to the consistency and dependability of a measuring instrument; using it repeatedly should give us the same or similar results every time. In real life, in most cases, we cannot use the same test over and over and expect the same results. Therefore, there are several practical ways that can be used to assess the reliability of a newly developed instrument. Four of these ways are presented in the chapter: *test-retest*, *alternate forms*, *internal consistency*, and *inter-rater reliability*. As you will see, the *correlation coefficient r* (see chapter 7) is also used in assessing test reliability. Keep in mind, though, that it is the responsibility of the test *developer* (rather than the test *user*) to conduct studies to demonstrate the reliability of the measuring tool.

You may be familiar with the way standardized test reports show bands of scores for students on each part of the test. The band relates to a concept called *standard error of measurement*, which is also discussed in this chapter. This is predicated on the idea that no test is 100 percent reliable, so the *true* score of test-takers can only be *estimated* based on their performance on the test. The chapter also explains various factors that affect reliability and the question of the desired level of reliability.

The term *reliable*, when used to describe a person, usually means that this person is dependable and consistent. Similarly, a reliable measure is expected to provide consistent and accurate results. If we use a reliable measure over and over again to measure physical traits, the same or very similar results should be obtained each time. For example, when we repeatedly use a precise scale to measure weight, we are likely to obtain the same weight time after time. When dealing with the affective domain (for example, self-concept or motivation) or even with the cognitive domain (for example, academic achievement), the performance of individuals on a measuring tool tends to change and is much less consistent. Factors such as moods, pressure, fatigue, anxiety, and guessing all tend to affect performance. Hence, even with a reliable measure it is hard to achieve a high level of consistency between measures. Since this book is intended for educators, our discussion of reliability focuses on procedures used in education.

WHAT IS RELIABILITY?

Reliability refers to the level of consistency of an instrument and the degree to which the same results are obtained when the instrument is used repeatedly with the same individuals or groups. This consistency may be determined by using the same measure twice, administering two equivalent forms of the measure, or using a series of items designed to measure similar concepts and traits.

The symbol used to indicate the reliability level is *r*, the same as that used for Pearson product-moment correlation coefficient (see chapter 7). As is explained later in this chapter, several procedures to assess reliability use correlation, so it is not surprising that the two share the same symbol. In theory, reliability can range from 0 to 1.00, but the reliability of measures of human traits and behaviors never quite reaches 1.00. Some very good achievement tests may reach .98, but probably not any higher than that.

It is the responsibility of the *test developer* to assess the reliability of the measure. Those who consider using the test must have information about the test reliability in order to make informed decisions. If those who use the instrument employ the measure to study groups and conditions that are similar to the groups and conditions used by the instrument developer, then the users of the instrument can assume that the measure has the same reliability as that reported by those who developed it.

UNDERSTANDING THE THEORY OF RELIABILITY

The classical theory of reliability states that an *observed* score, X (for example, a score obtained on an achievement test), contains two components: a *true score* (*T*) and an *error score* (*E*). The observed score can be described as:

$$X = T + E$$

The *true score* (*T*) reflects the *real* level of performance of a person, but it cannot be observed or measured directly. Based on the assumption that the *error scores* (*E*) are random and do not correlate with the true scores, the observed scores (X) are used to estimate the true scores. For some people, X is an *overestimate* of their true score; and for other people, X is an *underestimate* of their true score.

Theoretically, the true score of a person can be determined by administering the same measure over and over, recording the scores each time, and then averaging all the scores. In practice, though, people are tested with the same measure only once or twice at the most. Using the variance of the error scores and the variance of the observed scores, we can compute the reliability of the test using this formula:

$$\text{Reliability} = 1 - \frac{S_e^2}{S_x^2}$$

Where S_E^2 = Error variance
$\quad\quad S_X^2$ = Variance of the observed scores

As the formula shows, a decrease in the ratio of the two variances causes the reliability to increase. This ratio can be decreased either by decreasing the numerator or by increasing the denominator. The error component is related to the way the instrument is created and to the way it is administered. Consequently there are several ways to reduce the variance of the error scores (the numerator), such as writing good test items, including clear instructions, and creating a proper environment for the testing to take place. The variance of the observed scores (the denominator) can be increased by using heterogeneous groups of examinees (in terms of ability and performance) or by writing longer tests.

METHODS OF ASSESSING RELIABILITY

This chapter describes several methods to assess the reliability of tests: *test-retest*, *alternate forms*, and *internal consistency* methods. *Inter-rater* reliability is also discussed.

Test-Retest Reliability

The **test-retest reliability** is assessed by administering the same test *twice* to the same group of people. The scores of the examinees from the two testing sessions are correlated, and the correlation coefficient is used as the reliability index. The time interval between the two testing sessions is important and should be reported along with the reliability coefficient. When the interval between the testing sessions is short, the reliability is likely to be higher than in cases when the interval between the testing sessions is long.

The test-retest method of assessing reliability seems the most obvious approach because reliability is related to consistency over time. Nevertheless, there are several problems involved in this method of assessing reliability. First, people need to be tested twice, which may be time-consuming and expensive. Second, some memory or experience from the first test is likely to affect individuals' performance on the retest. Increasing the time interval between the two testing sessions may reduce this effect, but with a longer interval, new experiences and learning may occur and affect people's performance. Generally, it is recommended that the interval between retests not exceed six months.

Due to the problems associated with the test-retest method, this method is not considered a conclusive measure of reliability in education and psychology. It may be used, though, in combination with other methods designed to assess test reliability.

Alternate Forms Reliability

The **alternate forms reliability** is obtained when a group of examinees is administered the two forms, and the two sets of scores (from the two test forms) are correlated with each other. The two forms of the test should be equivalent in terms of their statistical properties (for example, equal means, variances, and item intercorrelation), as well as the content coverage and the types of items used. This method of assessing reliability is based on the assumption that if examinees are being tested twice, with two alternate forms of the same test, their scores on the two forms will be the same. As with test-retest, the correlation coefficient serves as the index of reliability.

There are two major problems involved in using this reliability assessment method. The first is that the examinees have to be tested twice, as was the problem with the test-retest method. The second problem is that it is very difficult, and often impractical, to develop an alternate form. If the purpose is merely to assess the reliability of a single test, then the alternate form method is unlikely to be used because it requires having a second form of the test. It should be noted that many commercial testing companies, especially those that develop achievement tests, construct alternate forms for other purposes. Thus, these forms can also be used to assess reliability. Alternate forms are useful for security reasons (for example, every other student gets the same form to reduce copying and cheating). They are also useful in some research studies, when one form is administered as a pretest and the other form as a posttest in order to eliminate the possible effect that previous exposure to the test may have on subsequent testing scores.

Measures of Internal Consistency

One major disadvantage of the two aforementioned reliability assessment methods is that the examinees have to be tested twice. **Internal consistency methods** to assess reliability allow the use of scores from a single testing session to estimate reliability. In essence, each item on a test can be viewed as a single measurement, and the test can be viewed as a series of repeated measures. There are several internal consistency methods that are based on the assumption that when a test measures a single basic concept, items correlate with each other and people who answer one item correctly are likely to correctly answer similar items. The reliability estimates obtained by internal consistency methods are usually similar to those obtained by correlating two alternate forms. The *split-half method*, KR-20 and KR-21, and *Cronbach's coefficient alpha* are some methods that can be used to estimate the test's internal consistency.

The Split-Half Method

In the **split-half method**, the test is split into two halves and the scores of the examinees on one half are correlated with their scores on the other half. Each half is considered an alternate form of the test. The most common way to split a test is to divide it into odd and even items, although other procedures that create two similar halves are also acceptable. However, dividing the test into the first half and the second half may create two halves that are not comparable. These two halves may differ in terms of content coverage, item difficulty, and students' level of fatigue and practice. To use the split-half approach, items on the test should be scored dichotomously, where a correct answer is assigned 1 point and a wrong answer is assigned 0 points.

The first step in the computation of the split-half reliability procedure is to obtain the scores from the two halves for each person. The scores from one half are then correlated with the scores from the other half. Unlike the first two methods discussed (test-retest and alternate forms), this correlation is *not* an accurate assessment of test reliability. In fact, it *underestimates* the reliability because it is computed for a test half as long as the actual test for which we wish to obtain reliability. Research has demonstrated that all things being equal, a longer test is more reliable. That is, if we have two tests with similar items, but one is shorter than the other, we can predict that the longer test is more reliable than the shorter test.

In order to calculate the reliability for a full-length test, the *Spearman-Brown prophecy formula* is used. This formula uses the reliability obtained for the half-length test to estimate the reliability of a full-length test. The Spearman-Brown prophecy formula is:

$$r_{full} = \frac{(2)(r_{half})}{1 + r_{half}}$$

Where r_{full} = Reliability for the whole test

r_{half} = Reliability for the half test (that is, the correlation of the two halves)

Suppose we want to estimate the reliability of a 30-item test, and the correlation of the odd-item half with the even-item half is $r = .50$. This correlation estimates the reliability for a 15-item test, whereas our test has 30 items. In order to estimate the reliability of the full-length test, the Spearman-Brown formula is applied as follows:

$$r_{full} = \frac{(2)(.50)}{1 + .50} = \frac{1.00}{1.50} = .67$$

In this example, the test developer should report the test split-half reliability as $r = .67$.

Kuder-Richardson Methods

G. F. Kuder and M. W. Richardson developed a series of formulas in an article published in 1937. Two of these formulas, **KR-20** and **KR-21**, are used today to measure agreement, or intercorrelation, among test items. As with the split-half method, these procedures can only be used for items that are scored dichotomously (right or wrong). KR-20 can be thought of as the average of all possible split-half coefficients obtained for a group of examinees. KR-21 is easier to compute, but it is appropriate only when the level of difficulty of all items is similar, a requirement that is not easily satisfied.

Cronbach's Coefficient Alpha

The *coefficient alpha* was developed by Lee Cronbach in 1951. **Cronbach's coefficient alpha** yields results similar to KR-20 when used with dichotomous items. Coefficient alpha can be used for tests with various item formats. For example, it can be applied to instruments that use a Likert scale, where each item may be scored on a scale of 1 to 5. Coefficient alpha measures how well items or variables that measure a similar trait or concept correlate with each other and it is considered by researchers to provide good reliability estimates in most situations. Readers of educational and psychological research are likely to see the coefficient alpha being reported as an index of reliability because it is a popular choice among researchers.

Inter-Rater Reliability

Inter-rater reliability refers to the degree of consistency and agreement between scores assigned by two or more raters or observers who judge or grade the same performance or behavior. For example, the process of scoring essay tests or observing and rating behaviors calls for subjective decisions on the part of those who have to grade the tests or rate the behaviors. Essays may be assessed using rubrics that include criteria such as content, organization, syntax and grammar, completeness, and originality. To assess the reliability of the essay-scoring process and the criteria used for grading, the essays first are read by two or more readers who assign a score on each criterion using a rating scale. The scores assigned by the scorers on the different criteria are then evaluated to see if they are consistent.

The scores from two or more essay readers can be used in two ways: (a) to compute a correlation coefficient or (b) to compute the percentage of agreement. The correlation coefficient and the percentage of agreement indicate the reliability and the consistency of the measure as used by the judges. A high correlation coefficient shows consistency between the readers. By providing clear guidelines for scoring, as well as good training, it is possible to increase the inter-scorer reliability and agreement. Similarly, when two or more observers rate certain behaviors using a rating scale, their ratings can be used to assess the reliability of the observation tool.

THE STANDARD ERROR OF MEASUREMENT

The reliability and accuracy of a test can be expressed in terms of the *standard error of measurement (SEM)*. The **standard error of measurement** provides information about the variability of a person's scores obtained upon repeated administration of a test. The standard error of measurement is especially suitable for the interpretation of individual scores. Since measures of human traits and behaviors contain an error component, any score obtained by such a measure is not a completely accurate representation of the person's true performance; the standard error of measurement allows us to estimate the range of scores wherein the true score lies. Tests that are more reliable contain a smaller error component than do tests that are less reliable. The reliability and the standard deviation (SD) of the instrument are used to compute SEM. The computation formula is:

$$SEM = SD\sqrt{1 - Reliability}$$

To illustrate how the SEM is computed, let's look at a numerical example. Suppose an achievement test has a standard deviation (SD) of 10, and a reliability of .91. The test's SEM is computed as:

$$SEM = 10\sqrt{1 - .91} = 10\sqrt{0.09} = (10)(0.3) = 3$$

Relating SEM to the normal curve model, we can state that 68 percent of the time the examinees' *true* scores would lie within ±1 SEM of their *observed* scores, and 95 percent of the time the examinees' true scores would lie within ±2 SEM of their observed scores. (See chapter 6 for a discussion of the normal curve.) For example, when a student obtains a score of 80 on this achievement test, 68 percent of the time the student's true score would be expected to be up to 3 points above or below the observed score of 80, or between 77 and 83. We can also predict that 95 percent of the time the student's true score will lie between 74 and 86, a range that is within 6 points of the obtained score (that is, within ±2 SEM). Clearly, it is desirable to have a small SEM because then the band of estimate (the range within which the true score lies) is narrower and the true score is closer to the observed score.

If you inspect the formula for the computation of SEM, you would realize that the reliability of a test affects its SEM. A lower reliability results in a higher SEM; therefore, there is a wider, less precise band of estimate. Assume that the reliability of the test in our previous example had been .64 instead of .91, SEM would then be computed as:

$$SEM = 10\sqrt{1 - .64} = 10\sqrt{0.36} = (10)(0.6) = 6$$

When the SEM is 6, it means that 68 percent of the time the student's true score would have been up to 6 points above or below the student's observed score of 80 (that is, between 74 and 86), and 95 percent of the time, the true score of these students would have been between 68 and 92. It is as if we are saying that although the student obtained a score of 80, we are 68 percent sure that the true score is somewhere between 6 points above to 6 points below that score.

FACTORS AFFECTING RELIABILITY

Heterogeneity of the Group

When the group used to derive the reliability estimate is *heterogeneous* with regard to the characteristic being measured (for example, typing speed, achievement level, or attitudes toward corporal punishment), the variability of the test scores is higher and, consequently, the reliability is expected to be higher. Test manuals that report the test's reliability are likely to include information on the groups used to assess reliability. Suppose, for example, that the group used to generate the test's reliability levels included students from grades 3 through 5. If the test is to be used with third graders only, the reliability of the test for the third graders is probably lower than that reported in the manual.

Instrument Length

As was mentioned before, all things being equal, a longer instrument is more reliable than a shorter instrument. In a shorter instrument, the probability of guessing the right answers is higher than in a long instrument. For that reason, creating a longer test can help provide a more stable estimate of the student's performance. The split-half reliability, which uses the Spearman-Brown formula, demonstrates the effect of the test length on reliability. It shows that a full-length test is more reliable than a test that is half as long. Hence, if you check the manual of a commercial test, you will see that the reliability levels of subsections of the test are usually lower than the reliability level of the whole test. In determining the desired length of any given test, though, it may be necessary to consider other variables, such as time constraints or the ages of the prospective students who will be taking the test.

Difficulty of Items

Tests that are too easy or too difficult tend to have lower reliability because they produce little variability among the scores obtained by the examinees. Tests where most of the items have an average level of difficulty tend to have higher reliability than tests where the majority of the items are very hard or very easy.

Quality of Items

Improving the quality of items increases an instrument's reliability. The process starts by writing clear, unambiguous items, providing good instructions for those administering and taking the test, and standardizing the administration and scoring procedures. Ideally, the instrument can then be field tested with a group similar to the one projected to take the test in the future. An item analysis can be performed to reveal weaknesses in the items and to help improve the test by reducing the error variance.

HOW HIGH SHOULD THE RELIABILITY BE?

Usually, self-made instruments, such as those created by classroom teachers, tend to have lower reliability levels than tests prepared by commercial companies or by professional test writers. Teachers and other practitioners may not have the time or the expertise to construct the tests, and they may not perform an item analysis or revise the items where needed.

Another point to keep in mind is that tests that measure the *affective* domain tend to have lower reliability levels than tests that measure the *cognitive* domain. The main reason for this phenomenon is that the affective domain behavior is less consistent than the cognitive domain behavior.

As a rule, important decisions, such as admitting students into a program, should not be based on a single test score because every test contains a certain level of error. Instead, multiple measures should be used for making important decisions. Batteries of achievement tests should report the reliability levels for the subtests, as well as for the total test. Additionally, the standard error of measurement (SEM) should be reported, whenever possible, to indicate the test's margin of error.

Decisions about the acceptable level of reliability depend to a great extent on the intended use of the test results. In exploratory research, even a modest reliability of .50 to .60 is acceptable (although a higher reliability is always preferable). For group decisions, reliability levels in the .60s may be acceptable. For example, in experimental studies that involve a comparison of experimental and control groups, individuals are not usually compared; rather, group information (for example, mean scores) is likely to be used for comparing the groups. When important decisions are made based on the results of the test, the reliability coefficients should be very high. Most commercial tests used for decisions regarding individuals have reliability levels of at least .90. Even though many classroom teachers do not have the time or the expertise to assess the reliability of the tests they construct, they should be aware of the issue of reliability in educational and psychological testing.

SUMMARY

1. **Reliability** refers to the consistency of a measurement obtained for the same persons upon repeated testing. A reliable measure yields the same or similar results every time it is used.
2. The affective and cognitive domains are more difficult to measure reliably than are physical traits.
3. The *real* level of performance for any individual, or the *true score* (*T*), cannot be observed directly. The observed score (*X*) is likely to *overestimate* or *underestimate* the true score for any given individual. This observed score equals the sum of the *true score* and the *error score* (*E*).

4. The reliability of a measure can be represented by this formula:

$$\text{Reliability} = 1 - \frac{S_e^2}{S_x^2}$$

5. Methods for *decreasing* the error component include writing good items, giving clear instructions, and providing an optimal environment for the test takers. Methods of *increasing* the variance of the observed scores include using heterogeneous groups of examinees and writing longer tests.

6. The reliability of a particular measure may be assessed using these methods: *test-retest*, *alternate forms*, and *internal consistency* approaches.

7. **Test-retest reliability** is assessed by administering the same test *twice* to the same group of people. The scores of the examinees from the two testing sessions are correlated, and the correlation coefficient is used as the reliability index.

8. The **alternate forms reliability** is obtained when a group of examinees is administered two alternative forms of the test and their two scores are correlated with each other. The correlation between the two alternate forms serves as the index of reliability.

9. **Internal consistency methods** to assess reliability use the scores from a single testing session. In these methods, each individual item becomes a single measurement, while the test as a whole is viewed as a series of repeated measures. Internal consistency methods include the *split-half, Kuder-Richardson methods (KR-20* and *KR-21)*, and *Cronbach's coefficient alpha*.

10. In the **split-half method**, the test is split into two halves and the scores of the examinees on one half are correlated with their scores on the other half. Each half is considered an alternate form of the test. *Spearman Brown prophecy formula* is then applied to compute the reliability level of the full-length test.

11. The reliability of an instrument can be assessed using the *KR-20* and *KR-21* formulas that are used to measure agreement, or intercorrelation, among test items. Scores obtained from a group of people who have taken the test one time can be used to obtain this reliability estimate.

12. **Cronbach's coefficient alpha** can be used to assess the reliability of instruments with different types of item formats using scores obtained from a single testing of the instrument.

13. **Inter-rater reliability** refers to the degree of consistency and agreement between scores obtained by two or more raters or observers who judge or grade the same performance or behavior.

14. The **standard error of measurement (SEM)** is used to assess the reliability and accuracy of the test in relation to its ability to accurately estimate the range of scores within which the true score lies. SEM is calculated using this formula:

$$\text{SEM} = \text{SD}\sqrt{1 - \text{Reliability}}$$

15. A smaller SEM allows for a more accurate estimate of the true score and, thus, provides a more reliable measure. Tests with higher levels of reliability have lower standard errors of measurement than less reliable tests.

16. Using the normal curve, we can state that 68 percent of the time the examinees' *true* scores would lie within ±1 SEM of their *observed* scores, and 95 percent of the time the examinees' true scores would lie within ±2 SEM of their observed scores.

17. Factors such as the *heterogeneity* of the group, the test *length*, and the *difficulty* and *quality* of the items affect the reliability of the measure.

18. Teacher-made tests tend to have lower reliability levels compared with tests constructed by experts.

19. Tests that measure the *affective* domain tend to have lower reliability levels than tests that measure the *cognitive* domain because affective domain behavior is less consistent than cognitive domain behavior.

20. Important decisions should not be made using a score from a single test because each test contains a certain level of error. Instead, scores from multiple measures should be used.

21. Decisions about the acceptable level of reliability depend to a great extent on the intended use of the test results.

CHECK YOUR UNDERSTANDING

1. Explain the concept of *reliability* and its importance in educational assessment.
2. Select an educational assessment tool that is used in your educational setting.
   a. What is its purpose and how is it used?
   b. How were the *reliability* and *validity* assessed and reported by the test constructor?
   c. What is your *personal evaluation* of this assessment tool? (If necessary, interview other colleagues who have more experience with this assessment).
3. What is the relationship between a test's *reliability* and *standard error of the measurement* (SEM)? Explain.
4. How high is the *reliability* expected to be for assessment tools that are used in education and psychology? Explain.

14

Validity

In addition to being reliable, a measuring instrument, such as a test, has to be valid. In chapter 14, we introduce to you the concept of *validity* and different ways to assess it. Validity refers to the extent to which a test measures what it is supposed to measure and the appropriateness of the ways it is used and interpreted. Several major types of validity are discussed in this chapter: *content*, *criterion-related*, *construct*, and *face* validity. An explanation of the process of assessing validity and the issue of *test bias* are also addressed in this chapter. As you will see, correlation is used in assessing criterion-related validity, whereas the other types of validity are evaluated by gathering various types of evidence.

As a classroom teacher or other educational professional, you may be involved in decisions regarding the choice of commercial educational tests. The notion of the validity of these assessments and how well they fit with your teaching and curriculum are of utmost importance. In this chapter, we further explain the concept of validity and its role in educational assessment.

WHAT IS VALIDITY?

The **validity** of a test refers to the degree to which an instrument measures what it is supposed to measure and the appropriateness of specific inferences and interpretations made using the test scores. It is not sufficient to say that a test is "valid"; rather, the *intended use* of the test should also be indicated. Keep in mind that validity is not inherent in the instrument itself and that an instrument is considered valid for a particular purpose only. For example, a test that is a valid measure of reading comprehension for students in the third grade is not valid as a measure of spelling for fifth grade students. Validation of a test involves conducting empirical studies where data are collected to establish the instrument's validity. A valid test is assumed to be reliable and consistent, but a reliable test may be valid only for a specific purpose.

There are several types of validity, including *content* validity, *criterion-related* validity, and *construct* validity. *Face* validity is also important and is included in this chapter.

CONTENT VALIDITY

Content validity describes how well an instrument measures a representative sample of behaviors and content domain about which inferences are to be made. In order to establish the content validity of a test, its items are examined and compared to the content of the unit to be tested, or to the behaviors and skills to be measured.

It is most important to assess the content validity of achievement tests. The achievement test developer should ensure that the items are an adequate sample of the content to be tested. If instructional objectives are available, the teacher may choose to examine the match between the test items and the objectives. Well-defined content domain and behaviors help increase the test's content validity.

The commercially available norm-referenced standardized achievement tests that are used by many school districts are designed to measure various standards, skills, and topics in the subject areas being tested. The developers of such tests usually provide information about the content being tested and show the match between the test items and the content they are designed to measure.

Educators who are assigned the responsibility of choosing a series of standardized, commercial achievement tests for their schools need to compare the items on the tests to their curricula and make sure they match. Because curricula are likely to differ from school to school, a particular standardized test may have high content validity for some schools but low content validity for other schools.

Teachers who write their own achievement tests should make sure that items on the test correspond to what was covered in their classes in terms of content, behaviors, and skills. For example, a teacher who teaches a unit on the Civil War and emphasizes understanding reasons and processes should not write test items that ask students to *recall* dates, events, and names, because such items lower the validity of the test.

CRITERION-RELATED VALIDITY

The process of assessing the **criterion-related validity** of a measure involves collecting evidence to determine the degree to which the performance on a measuring instrument is related to the performance on some other external measure. The external measure is labeled as the *criterion*. As part of the process to assess the criterion-related validity of the instrument, test developers can correlate it with an appropriate criterion. The correlation coefficient is used as the *validity coefficient*, and it is used to indicate the strength of the relationship between the instrument and the criterion. There are two types of criterion-related validity: *concurrent* validity and *predictive* validity.

Concurrent Validity

Concurrent validity is concerned with the evaluation of how well the test we wish to validate correlates with another well-established instrument that measures the same thing. The well-established instrument is designated as the *criterion*. For example, a newly created short version of a well-established test may be correlated with the full-length test. If the correlation between the two measures is high, it may indicate that they measure similar characteristics, skills, or traits. In order to establish concurrent validity, the two measures are administered to the same group of people, and the scores on the two measures are correlated. The correlation coefficient serves as an index of concurrent validity.

To illustrate, suppose a researcher develops a new IQ test that takes 30 minutes to administer and 20 minutes to score. This is much faster than the commonly used IQ tests. In order to establish the concurrent validity of the new IQ test, the researcher may correlate it with a well-established IQ test by administering both tests to the same group of people. A high positive correlation of the new test with the established IQ test would lend support to the validity of the new test.

Predictive Validity

Predictive validity describes how well a test predicts some future performance. This type of validity is especially useful for aptitude and readiness tests that are designed to predict some future performance. The test to be validated is the *predictor* (for example, the Scholastic Aptitude Test or the ACT test) and the future performance is the *criterion* (for example, GPA of college freshmen). Data are collected for the same group of people on both the predictor and the criterion, and the scores on the two measures are correlated to obtain the validity coefficient. Unlike concurrent validity where both instruments are administered at about the same time, predictive validity involves administering the predictor first, while the criterion is administered later in the future.

Suppose a researcher wants to establish the predictive validity of a music aptitude test for elementary school children. Forty third graders are administered the music aptitude test and are then given musical instruments in their schools. At the end of the year, the music teacher is asked to rate each student's musical achievement using a scale of 1 (poor) to 10 (excellent). Next, the aptitude scores are correlated with the teacher's ratings. A high positive correlation indicates that the aptitude test has a high predictive validity because it predicted accurately the students' end-of-year achievement in music.

You should keep in mind that tests that are intended to predict future performance may provide incomplete information about the criterion. For example, the music aptitude test may not always predict how well a student plays a musical instrument a year later. The reason is that this test may measure natural aptitude but probably not other factors such as motivation, perseverance, hours of practice, quality of music instruction, and parental support.

CONSTRUCT VALIDITY

The term *construct* is used to describe characteristics that cannot be measured directly, such as intelligence, sociability, and aggression. **Construct validity** is the extent to which an instrument measures and provides accurate information about a theoretical trait or characteristic. The process of establishing the instrument's construct validity can be quite complicated. The process includes administering the instrument to be validated to a group of people and then collecting other pieces of related data for these same individuals.

Suppose, for example, that a new scale has been developed in order to measure test anxiety. To demonstrate that the scale indeed measures test anxiety, the researcher first administers the scale to a group of people and then collects additional information about them. Those who score low on the test anxiety measure are considered to have a low level of test anxiety and are expected to exhibit behaviors and responses that are consistent with low anxiety levels. Conversely, those who score high on the test are expected to behave in ways that are compatible with a high level of test anxiety. Thus, establishing construct validity consists of accumulating supporting evidence. Evidence for construct validity is not gathered just once for one sample; rather, it is collected with the use of many samples and multiple sources of data.

FACE VALIDITY

Face validity refers to the extent to which an instrument *appears* to measure what it is intended to measure. The degree to which an instrument appears valid to the examinees and to other people involved in the testing process may determine how well the instrument is accepted and used. Additionally, face validity helps to keep test takers motivated and interested because they can see the relevance of the test to the perceived task. For example, a test with a high face validity that is used to screen a pool of applicants for certain positions is quite defensible as an appropriate instrument because applicants can see the test as relevant and perceive it as an appropriate measure.

Face validity is likely to be assessed based on a superficial inspection of an instrument. However, this inspection is not sufficient. The mere *appearance* of face validity is not a guarantee that an instrument is *valid* and that it truly measures what it is supposed to measure. You should be aware of the fact that face validity is not always found in discussions of validity, and it may not be considered by all to be as important as the other types of validity.

ASSESSING VALIDITY

Although we have identified several different types of validity, they are not necessarily separate or independent of each other. Establishing a measure's validity usually involves a series of steps of gathering data. Information provided by the instrument developer about its validity should include a description of the sample used in the validation process. Ideally, the characteristics of this sample are similar to those of future test takers.

Assessing the content validity of an instrument does not involve numerical calculation. Rather, it is a process of examining the instrument in relation to the content it is supposed to measure. In measuring criterion-related validity, the validity coefficient is used to describe the correlation between an instrument and a criterion. To be useful, the criterion has to be reliable and appropriate. The process of establishing the construct validity of an instrument includes the use of statistical methods, for example correlation) as well as procedures for gathering and comparing various measures.

TEST BIAS

Standardized tests, especially those used for admission, placement, and grading, are viewed at times as being *biased* against one group or another. A test is considered biased if it consistently and unfairly discriminates against a particular group of people who take the test. For example, certain tests are said to be gender-biased, usually discriminating against female examinees. Other tests may be considered biased against certain racial or cultural groups.

Norm-referenced tests are constructed to differentiate *among* examinees of diverse ability levels. This type of differentiation is not to be confused with the notion of test bias, where a test systematically discriminates against a particular group of examinees.

SUMMARY

1. **Validity** refers to the degree to which an instrument measures what it is supposed to measure and the appropriateness of specific inferences and interpretations made using the test scores. The intended use of the instrument should be indicated because it is considered valid for a particular purpose only.
2. The three basic types of validity are *content* validity, *criterion-related* validity, and *construct* validity.
3. **Content validity** refers to the adequacy with which an instrument measures a representative sample of behaviors and content domain about which inferences are to be made. In order to establish the content validity of the instrument, its items are examined and compared to the content of the unit to be tested, or to the behaviors and skills to be measured.
4. Instruments have **criterion-related validity** with respect to the relationship of scores on two separate measures. One measure is the newly developed instrument, and the other measure serves as a criterion. The criterion is usually a well-established assessment tool. There are two types of criterion-related validity: *concurrent* and *predictive*.
5. The correlation coefficient between the instrument and the criterion is called the *validity coefficient*, and it indicates the strength of the relationship between the two measures.
6. Assessing the instrument's **concurrent validity** involves evaluating the degree to which the results from the instrument correlate with another well-established instrument that measures the same thing.
7. An instrument has **predictive validity** if it can successfully predict future performance in a given area. The newly developed instrument is called the *predictor*, and the future performance is the *criterion*.
8. **Construct validity** refers to the extent to which an instrument measures and provides information about a theoretical trait or characteristic. To establish the construct validity of an instrument, it is necessary to collect additional data over a period of time and to correlate these data with the instrument results.
9. **Face validity**, which is not always recognized as a formal type of validity, refers to the extent to which an instrument *appears* to measure what it is intended to measure.
10. An instrument is considered *biased* if it systematically discriminates against a particular group of examinees.

CHECK YOUR UNDERSTANDING

1. Locate at least one *commercial test* that is given in your educational setting to assess students' academic performance. Inquire about its *validity* and *reliability* and summarize your findings, including your opinion about its use.

2. Locate at least one *psychological test* that is given to individuals or groups in your educational setting. Inquire about its *validity* and *reliability* and summarize your findings, including your opinion about its use.

3. Describe the process of how classroom teachers might be able to assess the *content validity* of tests or other assessment tools that they create and administer to their students.

4. Locate a report about test bias (any type of test). Briefly describe the main points made by the report's author, followed by your own opinion about this issue.

VII

PUTTING IT ALL TOGETHER

Choosing the Right Statistical Test

To help you plan your own research, chapter 15 provides you with an opportunity to apply the knowledge you have gained in the book about the appropriate use of statistical tests in different research situations. This chapter includes fourteen *hypothetical scenarios*, and you have to choose from a list of ten *statistical tests* that were introduced in the book. Your task is to decide which statistical test should be used to analyze the data described in the scenario in order to answer the research question. The chapter begins with a decision flowchart to help you determine which statistical test to use for your analysis; you can use this chart as you go through the scenarios to select the appropriate test.

To help you choose the right test for each scenario, we have two examples that take you step-by-step through the process. You can follow the same process used in these examples as you read the fourteen scenarios and try to select the correct tests.

You are now ready to embark on your own journey as an educational practitioner-researcher!

After researchers collect their data, they have to decide how to analyze their data in order to answer the study's questions and test its hypotheses. This chapter provides an opportunity for you to practice this important skill—that of choosing the proper statistical test to analyze the data you collected.

CHOOSING A STATISTICAL TEST: A DECISION FLOWCHART

The decision flowchart displays the various statistical tests covered in this book. The first level of the flowchart lists measurement scales of data (see figure 15.1). There are two choices: (a) *nominal* and (b) *interval/ratio* scales. The second level of the flowchart displays the types of hypotheses that are tested in the study. There are two types of hypotheses: (a) hypotheses that measure *differences* between groups or sets of scores and (b) hypotheses that measure *association* between variables.

In general, statistical tests may be classified into those designed to test hypotheses of association and those designed to test hypotheses of difference. It may be easier for you to distinguish between these two types of tests if you remember the following: tests that are designed to measure association between variables can indicate the presence or absence of an association as well as indicate the degree (or extent) of such an association. For example, the Pearson correlation that is used to test hypotheses of association can also provide information about the degree of association between two paired variables. This is done through the use of the correlation coefficient *r*. Tests that are designed to measure differences can also indicate the presence of a relationship between independent and dependent variables, but these tests cannot indicate the degree of the relationship. For example, a *t* test for independent samples may be used to measure the relationship between, let's say, gender and attitude toward homework, but it cannot *quantify* the magnitude of these relationships.

The third level of the flowchart asks you to decide whether the groups or variables in the study are *independent* or *paired* and whether there are *one* or *more groups* or *variables*. The final level in the flowchart includes a series of circles that list the *various statistical tests* that are introduced in this book.

After reading each scenario, decide which statistical test should be used to analyze the data and answer the research question that is stated or implied in that scenario. You can check your answers with those provided at the end of this chapter.

You can use the flowchart to assist you in selecting the proper statistical test. To use the flowchart, first determine the *scale* of measurement of the data in that scenario. Next, decide whether the research question or hypothesis in that scenario predicts a *difference* or *association*. Finally, decide whether there are one or more groups or variables in the study and whether they are *independent* or *paired*.

You may find that diagramming the design is often helpful as well. The answers to all of these questions should help you choose the right statistical test.

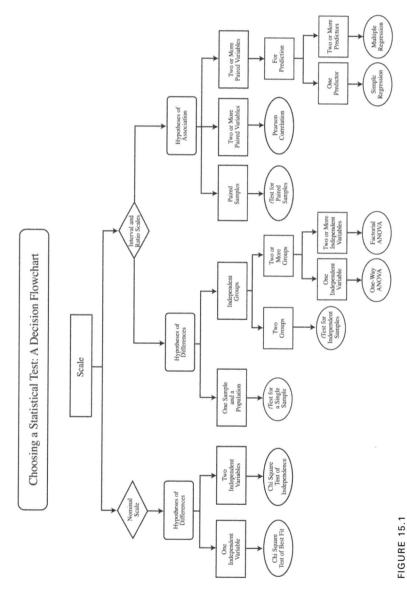

FIGURE 15. 1

Choosing a statistical test using a decision flowchart

In deciding which statistical test to use to answer the research questions and to analyze the data in the scenarios that follow, choose from these statistical tests:

1. Pearson correlation
2. *t* test for *independent samples*
3. *t* test for *paired samples*
4. *t* test for a *single sample*
5. One-way ANOVA
6. Two-way ANOVA
7. Chi square *test of best fit*
8. Chi square *test of independence*
9. Simple regression
10. Multiple regression

In order to help you get the "hang of it," start by reading the two examples that follow. The examples include answers to the questions about the scale of the data, the stated or implied hypothesis in the study, the groups used in the study, and some hypothetical data. As you work through the fourteen passages, we suggest that you ask yourself the same questions as those in the two practice exercises and create some hypothetical data points.

EXAMPLES

Example 1

Many studies comparing cooperative learning to traditional teaching have found that the social self-concept of students is improved when cooperative learning is used. A fourth grade teacher implements cooperative learning, and another fourth grade teacher continues with the traditional approach. At the end of the school year, a survey designed to measure social self-concept is administered to the fourth grade students in the two classes. The survey includes twenty Likert-scale items, with response choices ranging from *Strongly Agree* to *Strongly Disagree*. Scores on the self-concept measure can range from 20 to 100. The teachers want to compare the social self-concept scores of the students in both classes to see if those in the cooperative learning class have higher scores than those in the class using a traditional approach.

Answer:

1. *Scale*: *Interval*. The measure used in the study is a survey designed to assess the social self-concept of fourth grade students. Scores on the survey can range from 20–100.

2. In the survey, Likert-scale response options are provided, and such rating scales are measured on interval scales.

3. *Hypothesis*: A hypothesis of *difference*. The teachers predict that the social self-concept scores of students in the cooperative learning class would be significantly higher than the social self-concept scores of the students in the traditional classroom.

4. *Groups*: There are two groups of fourth grade classes that are independent of each other.

5. What might the data look like? See table 15.1.

Table 15.1. Solution to Example 1

Cooperative Learning	Traditional
45	34
50	29
38	41
Mean = 44.33	Mean = 34.66

Solution: Since the means of two independent groups are being compared, use the *t* test for *independent samples*.

Example 2

In many states in the United States, a portion of the residents' property tax is used to pay for public education. School board members all over the country have noticed that district residents who have school-age children attending the district's schools are more likely than other residents to support a tax increase to improve education. In a suburban school district, a nonbinding referendum about raising taxes to pay for education in the district is put on the ballot. The voters are asked to indicate their support or opposition to the referendum by marking the ballot with a *Yes* or *No* vote. The voters are also asked to indicate whether they have school-age children in the district's schools. The responses of the voters with and without school-age children are compared to determine whether those with children in the district's schools are more likely to support the tax increase compared with voters who do not have children in the district's schools.

Answer:

1. *Scale*: *Nominal.* The study compares two groups of voters and their responses. The response choices are *Yes* and *No*.

2. *Hypothesis*: A hypothesis of *difference*. The hypothesis predicts that residents *with* school-age children would be more supportive of the referendum compared with residents *without* school-age children in the district.

3. *Groups*: There are two groups that are independent of each other. One group is voters with school-age children in the district's schools, and the other group is voters without school-age children in the district's schools.

4. What might the data look like? See table 15.2.

Table 15.2. Solution to Example 2

| | Increase Taxes | |
Group	Yes	No
Have children in the district	87	13
Don't have children in the district	21	79

> **Solution:** Data are presented in a form of *frequencies*; therefore, use the *chi square test*. Since there are two variable (groups and response choices), use the *chi square test of independence*.

SCENARIOS

1. A junior high school principal wants to test whether students' achievement in one subject is related to their performance in other subjects. The principal selects a random sample of 200 students from her school and compares their standardized achievement test scores in mathematics, reading, language arts, and science to determine whether there is a relationship between students' scores in these four subjects.

2. Research has shown that students who learn keyboarding in the first or second grade tend to write and type longer essays compared with students who type using the "hunt and peck" approach. Students in the primary grades in one school in the district learn how to keyboard while no formal instruction is given to the primary grade students in another school in the same district. To assess whether formal keyboarding contributes to writing longer essays, the teachers of the students in both schools compare the length of an assigned essay (measured as number of words) written and typed by their students.

3. The director of food services in a school district is considering the addition of new items to the cafeteria menu. One of the new items is a green salad topped with strips of grilled chicken breast. After tasting the salad, students in the district's elementary, middle, and high schools are asked to indicate their preference by circling one of the following options: (a) *Add it to the menu*, (b) *Do not add it to the menu*, and (c) *No opinion*. The director of food services analyzes the data to determine if there are differences in the numbers of students in the district's elementary, middle, and high schools that chose each of the three response options.

4. A statistics instructor at a liberal arts college has noticed that psychology and sociology students seem to have more positive attitudes toward statistics compared with history and English students. The professor administers the *Statistics Attitudes Inventory* (*SAI*) scale to all students on the first day of the fall semester. The inventory contains twenty Likert-scale items with responses ranging from *Strongly Agree* to *Strongly Disagree*. The responses of psychology, sociology, English, and

history students are then compared to determine if there are significant differences in attitudes among the four groups of students.

5. A director of a large childcare center decides to train all of her teachers in the use of CPR. The Red Cross is invited to provide the training. To assess whether the CPR training program is effective, the Red Cross instructors administer a competency test to the teachers before and after the training program. The competency test contains multiple-choice questions and a performance assessment designed to measure CPR skills. The director and the instructors hypothesize that there will be a significant increase in *knowledge* of CPR on the posttest compared with the pretest scores on the *multiple-choice* test.

6. Many school districts administer readiness tests to students upon kindergarten entry. A publisher of a new kindergarten reading readiness test wants to convince potential users of the test that it can accurately predict students' academic performance in first grade. The test publisher offers to administer the reading readiness test, free of charge, to all kindergarten students in the district. After the test is administered, the reading readiness test score for each student is recorded. At the end of first grade, all students are administered a standardized achievement test. The reading readiness test scores obtained a year earlier and the scores on the standardized achievement test administered at the end of the first grade are studied to determine whether, as the reading readiness test publisher predicted, the reading readiness test can serve as a good predictor of end-of-year achievement of first grade students.

7. The faculty members in a large liberal arts college claim that the professors in the psychology department get paid more than faculty in other departments. The dean of the college assures the faculty that there is no significant difference in annual salaries between faculty members from different departments. The dean conducts a study to compare the annual salaries of the 32 psychology professors to the *mean* of the annual salaries of all the 530 faculty members in the college.

8. There are people who contend that spending hours each week in social networking online is a major contributor to a variety of academic problems, including low literacy rates. A group of middle school parents and teachers decides to investigate whether there is a relationship between the number of hours children spend on social network activities online and their grade point average (GPA). For two weeks, parents and their children record the number of hours the children spend in social networking online. For each student, the number of hours is recorded next to the student's grade point average (GPA). Teachers and parents can now analyze the data (number of hours spending on social networks online and GPA) and determine whether there is a relationship between these two variables.

9. The students in a high school mathematics class are learning about probability. They conduct an experiment with a four-sided spinner. The students hypothesize

that the spinner would land an equal number of times on each of the four sides. To test their hypothesis, the students spin the spinner 200 times and record the outcomes. Then they compare their observed results to those that are expected. Since there are four sides, the students expect the spinner to land 50 times on each side.

10. A high school physics teacher wants to evaluate two teaching methods in his classes that are studying about light. In one class, the teacher is using a textbook and demonstrations, while in the other class he is using inquiry-based experiments and investigation. At the end of the unit, the teacher gives a test to all of his students and compares the mean scores of the students in the two classes to determine which mean is higher.

11. Research to date has documented that there is a gender gap in computer use and in the field of Information Technology (IT). Some say that this disparity can be attributed in part to the fact that most electronic games are oriented toward boys' interests. A study is conducted with a randomly selected group of 250 fifth grade boys and 250 fifth grade girls who are given electronic games to play. Two electronic games are tested: Game A is an adventure game that requires competition among the players and game B is an adventure game that requires collaboration between the players. Half of the boys and half of the girls are given game A, and the other half is given game B. The researchers want to find out which computer game seems to appeal more to fifth graders and whether there are gender differences in preferences and attitudes toward the two games. A 20-item survey measuring attitudes and opinions is administered to the fifth graders after they play with their assigned game. Responses to each item on the questionnaire include four choices, ranging from *I liked it a lot* (4 points) to *I did not like it at all* (1 point). Scores on the survey items are added to create a total attitude score.

12. The directors of admissions for a large graduate school want to reexamine four variables currently used to select students for admission into the graduate programs. They want to determine whether these variables are good predictors of students' success in the program, as measured by the students' graduate program GPA. The variables that are used for selecting students are: (a) the Verbal score on the Graduate Record Examination (GRE), (b) the Quantitative score on the GRE, (c) the students' undergraduate GPA, and (d) the students' essay for admission. The records of 500 randomly selected students who have completed their graduate studies are used to test how well the four predictors predicted the students' graduate school GPA.

13. There are educators who claim that parents of younger children tend to be more satisfied with their children's schools compared with parents of older children. The teachers and administrators in a K–8 school disagree with this opinion and predict that there will not be significant differences in the attitudes of parents of students

in their school, regardless of the age of the students. To confirm their opinion, they examine results from a survey that is administered annually at the end of the school year to parents of students in kindergarten, third, and seventh grades. The survey is designed to measure the level of parents' satisfaction regarding the services, curricula, and programs provided by the school. The survey includes thirty questions, with responses to each question ranging from 1 (*Very Dissatisfied*) to 4 (*Very Satisfied*). A score of overall satisfaction is obtained for each of the respondents.

14. A sixth grade teacher has noticed that often when the students in his class talk about the television programs they had watched the evening before, the boys tend to discuss different programs from those discussed by the girls. To test whether there is a gender difference in the *type* of program watched by the children, the teacher asks all seventy-five sixth grade students in the school to list their favorite television program. The teacher then classifies the television programs into five categories: *action, drama, comedy, nature,* and *news*. The teacher compares the program types viewed by boys and girls to determine any gender differences.

ANSWERS

Scenario 1

The association between four measures obtained for the same group of people is assessed; therefore, *correlation* should be used. And, because the scale of measures is interval, use the *Pearson correlation*.

Scenario 2

The mean numbers of words on the essays typed by two independent groups are being compared; use the *t* test for independent samples.

Scenario 3

Data are presented in the form of frequencies; use the *chi square test*. Since there are two variables (grade level and response choices), use the *chi square test of independence*.

Scenario 4

The means of four independent groups (four college majors) are being compared; use the *one-way ANOVA*.

Scenario 5

Pretest and posttest means that are obtained for the same group of teachers are being compared; use the *t* test for paired samples.

Scenario 6

The reading readiness test is used to predict the first grade standardized achievement test for a group of students; use *simple regression*.

Scenario 7

A mean of one group (psychology department faculty) is being compared to the mean of the population (all other faculty members); use the *single-sample t test*.

Scenario 8

The association between two measures (number of hours children use social networking online and their GPA) obtained for the same group of people is assessed; use the *Pearson correlation*.

Scenario 9

Data are presented in the form of frequencies; use the *chi square test*. Since only one variable is used (the sides of a spinner) and the expected frequencies are of equal probability, use the *goodness of fit chi square with equal expected frequencies*.

Scenario 10

The means of two independent groups (two high school physics classes) are being compared; use the *t test for independent samples*.

Scenario 11

There are two independent variables (*gender* and *type of game*), each with two levels. Therefore, there are four independent groups in the study. The means of the four groups are compared; use the *two-way ANOVA*.

Scenario 12

All four measures are obtained for the same group of students. Three of these measures (GRE Verbal, GRE Quantitative, and undergraduate GPA) are used to predict the fourth measure (graduate GPA); use *multiple regression*.

Scenario 13

The means of three independent groups (parents of students in kindergarten, third, and seventh grades) are being compared; use the *one-way ANOVA*.

Scenario 14

The data are presented in a form of frequencies; therefore, use the chi square test. Since there are two variables (*gender* and the *type of television program*), use the *chi square test of independence*.

16

Planning and Conducting Research Studies

In chapter 16, we focus our attention on the process of *planning and conducting research studies*. We start with a brief discussion of *ethical* considerations in research. We then provide an explanation of the steps involved in planning your investigation and how to write a research proposal. While the specifics may differ and your proposal may be more or less formal than we describe, you would usually include an *introduction* where you explain the need for the study and present your research questions and hypotheses. A brief literature review is often included in the introduction to provide background and a framework for the proposed study. The next part is the *literature review*, where you present, summarize, and critique the research that relates to your topic, expanding on the information provided in the introduction. The third main part of the proposal is your proposed *methodology*. Here you address questions such as: Who will be studied? What tools and procedures will I use to collect my data? A list of all the *references* that were cited in the proposal is the last part.

The second major topic in this chapter is the *research report*. After you collect and analyze your data, you most likely will need to write a report to summarize and present your results. The first three parts of the research report (introduction, literature review, and methodology) are similar to the proposal but are more developed and provide additional details. You then add two sections, *results* and *discussion*. In the results section, you objectively present your findings, often with the aid of tables and charts. In your discussion, you interpret your results as they relate to your research questions and hypotheses and to other research on your topic.

Before researchers carry out studies, they need to plan and map out their steps. After the study is conducted, most researchers write a report that summarizes their findings. This chapter focuses on the process of writing research proposals for quantitative studies and on preparing reports that describe these studies. You may wish to refer to this chapter throughout the process of designing, conducting, and reporting on your research project. In this chapter, we focus on guidelines for students and researchers who conduct studies that examine quantitative data. Keep in mind, though, that your university or institution probably has its own specific set of rules and guidelines that may differ from those described in this chapter.

THE RESEARCH PROPOSAL AND REPORT

Both qualitative and quantitative studies require a clearly articulated research proposal prior to beginning the study. However, different research paradigms follow different guidelines and call for different approaches to the process of planning, conducting, and reporting research studies. While quantitative studies demand more detailed plans, proposals for conducting qualitative studies may be more tentative and less specific. Since this textbook is about quantitative statistics, the discussion here focuses on proposals and reports of quantitative studies.

Students who are writing proposals to meet degree requirements, such as theses or dissertations, probably need to follow specific guidelines given to them by their committees. Granting agencies are also likely to have their own guidelines for grant application proposals. Therefore, the discussion here is geared mainly toward students who plan to conduct research projects as a class assignment or for practitioners who would like to study their own settings.

Note that in discussing the different parts of the proposal and report in this book, the term *chapter* is used to describe the main parts of the proposal and report. The word *section* is used to denote a subpart of a chapter.

After a study is completed, it is described in a research report. Although the research proposal and the research report share common elements, they differ in several ways. For example, proposals include only three main chapters: *Introduction*, *Literature Review*, and *Methodology*. By comparison, research reports include these chapters *plus* two additional chapters: *Results* and *Discussion*. The three chapters of the proposal are also more developed and include further details in a research report (for example, exact description of the sample being studied; more complete literature review). Reports may also include an *Abstract* that summarizes the report and appears at the beginning. Another key difference between proposals and reports is the tense used. Proposals are written using *future* tense, whereas reports use *past* tense. Both proposals and reports also include a chapter called *References* which lists all the sources cited or quoted in the text. Additionally, an *Appendix* may be found in both proposals and reports.

When writing a proposal or report, you may be asked to use a particular writing style. The most well-known writing style, and the one used by most universities, is the one described in the *APA Publication Manual*.[1] Nonetheless, other styles may be used as well, and you should check to see which style you should follow.[2]

Before writing your research proposal, you should investigate your topic by reading about it as much as possible. By reviewing the literature you become well-informed about your topic, gather background information, learn about current trends and theories related to your topic, and identify gaps and controversies in the literature. All of these should help sharpen your focus and select your own specific research topic. The literature review process can also prevent you from unintentionally duplicating other studies and will help you avoid other researchers' mistakes as well as benefit from their experience.

There is a wealth of information available electronically on the Internet. The electronic data search techniques change and are updated at a rapid pace; it is probably a good idea for you to consult with your librarian in order to learn about the most recent techniques for electronic literature search. Your librarian can also help you obtain scholarly articles and help you stay away from studies that did not follow acceptable standards of research.

Every researcher undertaking a research study should be cognizant of ethical considerations involved in research. Before discussing the research proposal, we briefly review the ethics of research. This chapter then continues with a discussion of research proposals, followed by a discussion of research reports.

RESEARCH ETHICS

When planning your study, you should be concerned with *research ethics* and guarantee that the rights of those you study are protected at all times. As a researcher, you have to ensure that the participants are well-informed of the nature of the study and that you have not placed them in risky situations. Adult participants should provide their consent to participate in the study, while parents or legal guardians should provide consent for minors in their care. The study's participants also should be given an opportunity to withdraw from the study at any time. People's request for privacy should be honored and their confidentiality should be assured.

Ethical considerations are especially important in experimental studies, where participants undergo planned interventions. Practitioners who conduct research in their own settings should also maintain high ethical standards and be aware of all possible consequences of their studies. When you research your own practice, you should ensure the rights, welfare, and well-being of the study's participants. If you conduct

---

1. HINT: At the time this book was written, the most recent edition of the *APA Publication Manual* (published by the American Psychological Association) was the sixth edition, dated 2010. As you read this book, check to see if there is a newer edition. The most recent APA publication guidelines may also be found on the Internet at http://www.apastyle.org.

2. HINT: Another writing style that is used by university students and faculty is found in the *Chicago Manual of Style*, which was originally written by Turabian in 1937 and has been updated several times since then. The *MLA Handbook*, published by the Modern Language Association of America, is also used in some cases.

your investigation at your school, make sure you check with your supervisors before undertaking your study and secure their permission to conduct the study.

Several professional associations provide guidelines for their members regarding studies that involve human subjects.[3] Institutions, such as universities, as well as government offices and granting agencies, may request that all research proposals undergo a review by a human subjects review board as part of the proposal approval process. Many school districts also have research review boards to ensure that high ethical standards are maintained in their schools.

In any type of research that you may be involved in—whether you collect the data or share existing data with others—you should familiarize yourself with the recent policies. Following are three websites that you may want to visit. (New or updated sites may be available after the publication of this book, so make sure you check the Internet for updates.)

1. The Family Educational Rights and Privacy Act (FERPA): http://www.aera.net/ResearchPolicyAdvocacy/ResearchPolicyAdvocacy/FERPA/tabid/10271/Default.aspx
2. Revised FERPA regulations regarding sharing information about students: http://www.governor.wa.gov/oeo/reports/ferpa_2011_parent_overview.pdf
3. Protecting student privacy while using online educational services: http://www.ed.gov/news/press-releases/department-releases-new-guidance-protecting-student-privacy-while-using-online-e

THE RESEARCH PROPOSAL

After deciding on a topic for your study, your next step is to write a *research proposal*. A proposal may be viewed as the blueprint for the study. It provides a rationale for the study and an explanation of the reasons the study should be conducted. A well-written, carefully planned proposal helps you plan ahead, anticipate your needs, and outline a realistic timetable.

A typical proposal contains the following: *Introduction, Literature Review*, and *Methodology*. A list of sources used in the proposal is also included in the *References*. All proposals are expected to include these three parts, whether they are submitted to fulfill requirements for a degree or to request grants. In some cases, the *Introduction* and *Literature Review* are combined into one chapter, called *Introduction*. Regardless of the number of chapters required, all proposals should have an introductory chapter that includes a statement of purpose or research questions and any hypotheses, as well as a brief review of the literature. Additionally, a proposal should contain a description of the study's methodology and a list of the references cited. The *Methodology* chapter

---

3. HINT: Consult professional organizations such as the American Educational Research Association and American Psychological Association for their guidelines regarding ethical principles.

may also include a section entitled *Data Analysis*. Information about data analysis can be included either as a separate section in *Methodology* or as part of the *Procedures* section.

Next, we present a description of the main components of the research proposal: *Introduction, Literature Review*, and *Methodology*. A brief discussion of *References* is also included.

Introduction

The *Introduction* chapter introduces the study by stating the problem to be investigated, the purpose of the study, the rationale for conducting the study, and the study's potential contributions to the field. This chapter also includes the research questions and any hypotheses stated by the researchers prior to the study.[4] The background of the problem should be briefly developed in this chapter, but the main discussion of published background information should be included in the *Literature Review* chapter.

The *Introduction* chapter includes a *statement of the problem* that is written as a question or a declarative statement and is usually placed at the beginning. For example, a declarative statement may state that the study was designed to investigate the effect of inclusion on the social self-concept of included middle school children. An example of a problem stated as a question is: Are there differences in grade point averages (GPAs) between high school athletes and nonathletes?

The *Introduction* chapter should include a *rationale* and a clear explanation of the need for studying the issue and for finding solutions to the problem. After reading this explanation, the reader should be able to understand the potential contribution of the proposed research to educational practice or theory. For example, with regard to inclusion, it can be argued that those who promote it claim that it enhances the social skills of included children. Now, after many schools have had an opportunity to include children with special needs for a number of years, has inclusion lived up to its expectations? Do included middle school students have healthier self-concepts than similar students who are not included? The second topic dealt with high school athletes. There are people who have an image of high school athletes as "jocks" who are not very smart, take "easy" classes, and do not spend much of their time studying. Those who coach and work with athletes contend that this image is unjustified and that high school athletes perform academically as well as, or better than, their nonathletic peers. Thus, a study should be undertaken to compare the GPAs of high school athletes and nonathletes.

Many proposals also include research hypotheses, especially those that propose experimental studies. The exact placement of the hypotheses may differ. Most guidelines require placing them in the *Introduction* chapter, but others may suggest placing the hypotheses in the second or third chapter (*Literature Review* or *Methodology*) of the

---

4. HINT: See chapter 2 for a discussion of research questions and hypotheses.

proposal. We suggest that you check the specific guidelines given to you for further direction about the placement of the study's hypotheses.

The *Introduction* chapter also includes a brief review of selected sources that are most related to the topic. Those references are discussed in greater detail in the *Literature Review* chapter that follows the *Introduction*. Proposals may also include *definitions of key terms*, *assumptions*, and *limitations* of the study.

Literature Review

The *Literature Review* chapter summarizes literature related to the topic being investigated. In the proposal, the review of related literature tends to be limited in scope, citing briefly a small number of studies. Later, when writing the research report, this section is expanded. Literature reviews in dissertations and thesis proposals are expected to include the most important studies on the topic. In less formal proposals, such as those written as part of class research projects, the literature review is not likely to be comprehensive due to time constraints. When the literature review in a proposal is comprehensive and includes a discussion of a number of subtopics, we recommend that you consider the use of subheadings to better organize the topics you discuss. Adding a summary at the end of the literature review is also recommended, especially for longer reviews.

When writing the review, it should be organized by topic, rather than as an annotated bibliography or a series of summaries of articles, reports, or books. As the writer of the proposal, it is your responsibility to synthesize the research on your topic and to point out controversies in the field, as well as similarities, agreements, or disagreements among researchers who have conducted research on your topic. Additionally, existing gaps in knowledge and practice should be noted. You should also include a critique of studies you review and point to their shortcomings and contributions.

The information presented in the *Literature Review* should be properly attributed to its authors to avoid plagiarism. Sources must be acknowledged whether quoted directly or summarized. You should summarize key ideas, findings, and conclusions of other researchers. It is best to quote very little, if at all, and quote only phrases or ideas that are so well stated you feel you cannot summarize them accurately. Remember to include the page numbers for the quotes you use, when available. Try to keep the tone of your writing objective and unbiased and present a balanced discussion of all views, even those you may personally oppose.

A number of studies that are discussed in the *Literature Review* chapter would most likely be referred to again in the *Discussion* chapter of your research report. In that chapter, results from your own study should be examined and related to the existing body of knowledge in the field.

Methodology

The *Methodology* chapter in the proposal is designed to describe your plan of action and to clarify for the reader how you are going to investigate the research questions and test the hypotheses. The description of your methodology should be specific enough to communicate to the reader that you have carefully planned every step of your study.

When planning the specifics of the study, you should ask yourself questions such as:

- Is my study feasible?
- Do I have the expertise, resources, and know-how to carry out the study?
- Have I set a realistic timetable to design, conduct, and complete the study?
- Can I obtain the cooperation and collaboration of others, if needed?
- Do I need permission to conduct the study?
- What data collection instruments should I use?
- How can I recruit participants for my study?

The three main sections of this chapter are *Sample* (or *Participants*), *Instruments* (also called *Tests* or *Measures*, or *Data Collection Tools*), and *Procedure*.[5] Sections about *Design* or *Data Analysis* may also be included in this chapter.[6]

Sample

In the *Sample* section, describe those who will participate in your study. In most studies, your participants are likely to be people, but a sample can be comprised of a group of cases or items. You should present information related to the sample, such as how the sample will be selected, the size of the sample, and relevant demographic characteristics about the sample. You, as the researcher-author, have to decide which demographic characteristics are relevant to your study. For example, family income, age, or IQ scores of the study's participants may be considered important demographic characteristics in one study, but not in another.

Obviously, the *exact* information about the sample in your own study (for example, the mean age or the number of boys and girls in each group) may not be known until you actually conduct the study. Nevertheless, the *Sample* section should communicate your plans and intentions and provide a general description of the study's participants.

Instruments

The instruments you plan to use in the study should be clearly described and their purposes explained. If you plan to use existing instruments that were developed by others, their reliability and validity should be reported.[7] Additional information about

---

5. HINT: In some textbooks and journal articles, the term *subjects* is used in place of *sample* or *participants*. Note, though, that currently most guidelines recommend using the terms we use in this book, namely *sample* and *participants*, rather than *subjects*.

6. HINT: Some writing guidelines further divide the *Methodology* chapter (especially in experimental studies) into additional sections, such as *Materials*, *Independent Variables*, and *Dependent Variables*.

7. HINT: Several publications provide information about published tests. For example, the Buros Institute publishes the *Mental Measurement Yearbook* and *Tests in Print* (see http://www.unl.edu/ buros/); online test reviews are also available on this website. See also chapters 13 and 14 in this book for additional information about reliability and validity.

the instruments may also be reported when available. For example, you may describe the number and type of items used, the length of time required to complete the instrument, and how test norms are reported. Check for copyright information and for permission to use the instrument or to include it in your proposal.

If you plan to develop a data collection instrument (for example, a survey or an achievement test), explain how you plan to construct it and the type of items you will use. When appropriate, you should also discuss how you plan to assess the instrument's reliability and validity and whether you plan to pilot test it first before using it in your full-scale study. It is also advisable to include sample items of your proposed instrument. You can include the complete instrument in the proposal's *Appendix*.

Procedure

The *Procedure* (or *Data Collection Procedures*) section describes how the study will be conducted. It explains, in as much detail as possible, what will happen and how you will carry out the proposed investigation. This section is especially important in experimental studies that require a detailed description of the intervention. Examples of information to present in this section include a description of the training required to implement a new teaching method and the instructions to be provided to respondents who are asked to complete a survey. This section should also contain a realistic timetable for the different phases of the study.

Data Analysis

A description of your plans for organizing and analyzing the data to be collected in the study should be included in the *Data Analysis* section. It is a good idea, when possible, to decide in advance which statistical tests and techniques you will use to investigate your research questions and test your hypotheses.

In descriptive studies, the data analysis may include tables to summarize basic descriptive statistics, such as percentages, means, and ranges. Graphs and charts are also likely to be used in such studies. Other studies, especially those using inferential statistics, may require statistical techniques, such as the *t* test and analysis of variance (see chapters 9 and 10, respectively).

References

The last chapter in a proposal is *References*, where you list all the sources cited or quoted in the proposal. The exact way to list the references depends on the writing style used. For example, *APA Publication Manual* requires that you list all the references in alphabetical order according to the authors' last names. Other styles may specify that references be listed in the order they are cited in the text. The exact rules and guidelines for formatting each individual reference also vary depending on the writing convention used. Regardless, all sources cited in the text should be listed in the *References* chapter, and all references listed should have been cited in the text.

THE RESEARCH REPORT

After conducting your research study and analyzing the data collected, you are now ready to write your report. As with the proposal, specific guidelines for writing research reports may vary, depending on the nature and purpose of the report. For example, if you have to write the report as part of a class research project, your instructor may give you particular guidelines to follow. If this is a thesis or a dissertation study, your committee will ask you to follow your university's guidelines. In general, though, all quantitative research reports are likely to have similar components.

Research Reports usually have six chapters: the first five are *Introduction, Literature Review, Methodology, Results,* and *Discussion.* The sixth chapter, *References,* includes a list of sources cited or quoted in the first five chapters. And, as was the case with the proposal, the references listed in that chapter should correspond to those cited in the text. An *Appendix* is also likely to be included.

The first three chapters of a research report are the same as those found in a proposal (that is, *Introduction, Literature Review,* and *Methodology*), but they are longer, more detailed, and better developed. In a final report, the *Literature Review* chapter usually includes more references and citations than in a proposal. This is true especially in theses and dissertations, where the author is expected to include *all* the sources and literature related to the research topic. When writing the *Methodology* chapter in the report, you are likely to include more specific information about the study. For example, you can now report the exact number of those who participated in the study or how many responded to a questionnaire you administered. And, unlike during the stage of writing the proposal, now you have the results of the study and can write the *Results* and *Discussion* chapters.

At times, the *Introduction* and *Literature Review* chapters are combined into one chapter, titled *Introduction.*[8] For example, some journal editors advise authors who are interested in submitting manuscripts to their journals to combine the two chapters. Regardless, dissertations and theses guidelines typically advise students to keep the two chapters separate.

You may also be asked to write an abstract, which is usually found at the beginning of the report, before the introduction. The *Abstract* summarizes the study and focuses on the study's research problem, methodology, main results, and major conclusions. Abstracts are usually limited in length (that is, number of words), from about one short paragraph of 100 words to about one-and-a-half pages (or 1,000 words). Because the length of the abstract is limited, it has to be succinct and present only the most

---

8. HINT: According to *APA Publication Manual,* there is no need to type the word *Introduction* because its placement at the beginning of the manuscript identifies it as the introduction. It is always a good idea to check the specific typing guidelines given to you.

important points. If you plan to submit an article for publication, check the guidelines specified by the journal editors.

Many reports, especially theses and dissertations, include an *Appendix* at the end of the report. The appendix includes information that is too lengthy or too specific to be included in the text of the report. For instance, the appendix may include the complete survey used in a study or a letter to parents asking for their permission to observe their children.

Results

The *Results* chapter presents the study's findings. This chapter includes numbers, tables, and figures (for example, charts and graphs). The information presented and conveyed to the reader in this chapter should be written objectively, factually, and without expressing personal opinion. For example, you should not make statements such as, "We were disappointed to see that the anti-bullying program we implemented in our school did not significantly reduce the number of bullying incidences."

A good way to organize and discuss your findings in this chapter is to reiterate the hypotheses, one by one, and present the data that were collected to test each hypothesis. It is your decision as to what data to present in a narrative form and what to present in tables or figures. Although quite often, the tables and figures are accompanied by a narrative explanation, there is no need to describe in words *everything* that is presented in a numerical or visual form. Instead, "walk" the reader through the numerical and visual information. As the author, you should highlight the main findings, point to trends and patterns, and guide the reader through the information you present. For example, in a table displaying results from three independent-samples *t* tests, you can state that the second *t* value, which was used to test the second research hypothesis, was statistically significant at $p < .01$, and that the mean of the experimental group was 8 points higher than the mean of the control group. There is no need to repeat in the narrative *all* the numerical information reported in the tables. Or, suppose your *Results* chapter includes a double-bar graph that is used to show trends and differences in the percentages of male and female teachers who teach preschool, elementary school, and high school. You may explain that the trend is for the percentage of male teachers to increase with grade level, whereas the percentage of female teachers decreases from preschool to high school.

As to the actual typing of tables and figures, consult the guidelines given to you. Each style has different requirements, and those requirements can be quite specific. For example, according to APA style the title of a table should be typed *above* the table while the caption (that is, title) of a figure should be typed *below* the figure.

The tables and charts you construct should be easy to read and understand. In all likelihood, the computer printouts produced by the statistical program you use are not going to be "reader-friendly" and you will probably need to retype them following the guidelines given to you.

Discussion

Results from the study are discussed, explained, and interpreted in the *Discussion* chapter. The results are examined to determine whether the study's hypotheses were confirmed. This chapter allows you to offer your interpretation and explain the meaning of your results. If the findings are different from those that were predicted by the hypotheses, you have to provide tentative explanations for those discrepancies. For example, some possible explanations for unexpected results in a study may be that the sample size was too small, the study was too short, directions given to participants were not followed properly, the instruments were not valid or reliable, or the survey response rate was too low. Or, in some studies, one may speculate that the responses given by the participants were contrary to what was expected because people were dishonest in their responses or were reluctant to share certain sensitive information with others.

Often, the study's shortcomings are discussed in a section called *Limitations of the Study*. For example, you may explain that the results of the study should be generalized only to other groups with demographic characteristics similar to those of the study's participants. At other times, you may include as a limitation the fact that the study was too short.

Besides discussing the results from your own study, you should include in this chapter a discussion of your findings in relation to findings from other researchers. Point to examples where your own research supports or contradicts other researchers whose work was discussed in the *Introduction* and *Literature Review* chapters. By doing so, you demonstrate how your study relates to the field and contributes to its knowledge base.

Other sections, such as *Conclusions*, *Recommendations*, *Implications* (or *Implications for Practitioners*), and *Suggestions for Further* (or *Future*) *Research* may follow the discussion of the findings and be included in the *Discussion* chapter. Again, consult the guidelines given to you to find out what you are expected to include in the research report.

SUMMARY

1. Different research paradigms follow different guidelines and require different approaches to the process of planning, conducting, and reporting research studies.

2. *Quantitative* studies demand a more detailed research plan, compared to proposals for *qualitative* studies. This chapter discusses how to plan, conduct, and report *quantitative* research, with a focus on numerical data.

3. When writing a proposal or report, researchers usually follow a specific writing style, such as APA style.

4. Researchers, including those studying their own practice, should follow *ethical* principles. This is especially important in experimental studies where participants undergo a planned intervention. The rights of the study's participants should be protected at all times. Other guidelines are also discussed in the chapter.

5. The research *proposal* can be viewed as the blueprint for the study. It also provides a rationale for the study and an explanation of the reasons the study should be conducted.

6. Reading as much as possible about your topic will assist you in narrowing down and selecting your specific topic and in writing the literature review. It will also prevent you from unintentionally duplicating other studies and help you select methods and procedures for your study.

7. A research proposal includes the following chapters: *Introduction, Literature Review, Methodology,* and *References.*

8. The *Introduction* chapter of your proposal should provide a brief background of the problem and explain the significance of the topic and its potential contributions to the profession. This chapter should also include a rationale for your study in order to convince the reader that your topic is worthwhile.

9. Your proposal should include a *statement of the problem* that explains the question to be explored. It should be in the form of a declarative statement or a question.

10. Most proposals include *hypotheses.* This is true especially in proposals that involve experimental studies.

11. *Definitions, assumptions,* and *limitations* may also be included in a proposal.

12. The *Literature Review* chapter summarizes research related to the topic being investigated. All information presented in the review should be properly acknowledged and attributed to its authors to avoid plagiarism.

13. The *Methodology* chapter is designed to describe your plan of action and to clarify to the reader how you are going to answer the research questions and test the hypotheses.

14. Information about those who will participate in the study and their demographic characteristics is found in the *Methodology* chapter, under *Sample* (or *Participants*).

15. A clear description of the instruments to be used in the study should be included in the *Methodology* chapter under *Instruments* (also called *Tests* or *Measures,* or *Data Collection Tools*). When appropriate, sample items, as well as information about instruments' reliability and validity, should be included.

16. The *Methodology* chapter also includes a *Procedure* section that describes how the study will be conducted. This section is especially important to

include in experimental studies that require a detailed description of the planned intervention.

17. The *Data Analysis* section should describe how you plan to organize and analyze the data to be collected in the study.

18. The last chapter in a proposal is *References*. All sources cited or quoted in the text should be listed in this chapter.

19. Most *research reports* include the following chapters: *Introduction, Literature Review, Methodology, Results, Discussion*, and *References*.

20. The first three chapters of the report are the same as those in the proposal (that is, *Introduction, Literature Review*, and *Methodology*). However, these chapters in the report are longer, more detailed, and better developed than in the proposal.

21. An *Abstract*, summarizing the study, may also be included in a research report. It is usually placed right at the beginning of the report (before the *Introduction*).

22. The information in the *Results* chapter should be reported objectively, factually, and without expressing personal opinion. This chapter tends to be comprised of words, numbers, tables, charts, and figures. A good way to organize your findings is to reiterate the hypotheses (or research questions) one by one and present the data that were collected to test each hypothesis.

23. The results from the study are discussed, explained, and interpreted in the *Discussion* chapter. This chapter refers back to the study's research questions and hypotheses and discusses them. It also places the results from the study in relation to findings from previous studies.

24. The research report may also include *Conclusions, Recommendations*, or *Suggestions for Further Research*.

CHECK YOUR UNDERSTANDING

1. Describe an example within or outside of education where educators or researchers need to submit a *formal research proposal* prior to conducting a research study or when applying for a grant.

2. List three advantages of writing a research proposal.

3. Why do teachers have to consider *ethical aspects* when collecting data and conducting research *in their own classroom*? Explain.

4. Find a *published article* that reports *statistical* data and read it thoroughly.
   a. Do you find the article well written and easy to follow and understand?
   b. Is there a proper explanation of the background information and the need for the study?
   c. Do you think that the data are reported in a clear and logical way, and with proper explanations?
   d. Are the conclusions based on the study? Do they include practical implications for practitioners?

Using Statistical Tests to Analyze Survey Data

The focus of chapter 17 is on the use of statistical tests to analyze survey data. Surveys are used in many disciplines to collect a variety of data, both quantitative (numerical) and qualitative (narrative). In this chapter, we present some ways to analyze numerical data using statistical tests that were discussed in the book. We start this chapter with a brief overview of surveys and highlight a few points related to the planning, construction, and administration of surveys. We then provide several examples of surveys on various topics and show how statistical tests can be used to analyze data. We start with an illustration of a simple data summary that is presented in a frequency table, where tally marks are used to record the data. We then show examples of the following statistical tests to analyze survey data: Pearson correlation, *t* tests for independent samples and paired samples, one-way ANOVA, and the chi square test.

Surveys are used in all areas of our lives to collect a variety of information. In all likelihood, you have been asked at some point to respond to a survey. In your role as an educator, or in other areas of your life, you may have also had to construct a survey to gather information. In this chapter, we present examples to illustrate how you can apply statistical tests that were introduced earlier in this book to analyze survey data. Keep in mind that even though narrative, qualitative data can easily be collected using surveys, our discussion here focuses only on survey items that are used to gather quantitative data.[1]

WHAT ARE SURVEYS?

A **survey** is usually undertaken to gather information on a selected group of respondents about a topic that is of interest to the researchers.[2] In most cases, those surveyed are considered a sample of a target population that is the focus of the investigation. In other cases, those surveyed comprise the total group that is of interest and every individual in that group is included in the survey. Information obtained using surveys may include factual information (for example, highest level of education obtained); behavior (for example, average number of hours watching TV during the week); or attitudes and opinions (for example, preferences of video games, attitudes toward school, or level of satisfaction with an experience or service provided). The information gathered is then analyzed and summarized and conclusions are drawn regarding the issues being investigated.

Surveys are used in all walks of life, by individuals, private and public organizations, corporations, political groups, and the government. As a university student or professional educator, you are probably very familiar with surveys and have taken or administered them more than once. Most surveys are self-administered using a variety of delivery methods. Delivering surveys digitally makes administering surveys easier and more efficient. For example, surveys can be created by researchers using their own design and templates and sent to respondents via email; researchers can also create surveys using more professionally designed templates available through websites (such as SurveyMonkey and Google Docs).

Surveys can also be used in experimental studies that include an intervention. For example, a study was conducted to assess the effectiveness of a new media literacy program on teenagers' perceptions of smoking. Students who indicated that they were not opposed to smoking were divided into two groups: One group of students heard antismoking messages in a traditional class; the other group heard the same messages through a media literacy program. At the end of the program, a survey was administered to the students in both groups. The results showed that those in the second group (media literacy program) stated that they were less likely to smoke as compared with those in the traditional antismoking class).

---

1. HINT: Another point to keep in mind is that this chapter is not meant to be a comprehensive guide on how to conduct surveys; this topic is beyond the scope of this book.

2. HINT: Although some make a distinction between *surveys* and *questionnaires*, in this book these terms are used interchangeably.

We start the discussion here with a brief description of some popular item formats that can yield numerical data or be used to categorize respondents into groups. These items usually include a series of predetermined responses—either numerical or verbal—that are chosen by the respondents. We then offer some recommendations for survey planning and design which we suggest you consider before you embark on your study. Lastly, we show examples of several statistical tests that can be used to analyze survey data.

CONSTRUCTING SURVEYS

There are many types of survey items; each is designed to yield different types of data. Surveys may contain a variety of response choices, depending on the type of item and the information that the survey constructors want to gather. For example, some response choices are discrete (for example, Yes/No, Female/Male), some are on a continuum (for example, a 5-point Likert scale with response choices going from Strongly Disagree to Strongly Agree, or from Never to Always), while others ask respondents to rank-order a set of responses. Surveys may also include items that are called *contingency* or *filter* items that are intended to ensure that respondents are knowledgeable and informed about the questions or topics presented in the items that follow. For example, in a survey conducted by a school district about its electronic newsletter for parents, the first question might be about whether the parent reads the newsletter. This question is a contingency or filter question because those who do not read the newsletter should not continue to respond to questions specific to the content of the newsletter.

Before starting to write the survey items, consider the logistics of administering the survey and the information you want to obtain. Here are some suggestions for you to contemplate before beginning your study:

1. **Your target audience**: Who do you want to study? Do you plan to study everyone or select a sample?
2. **Access to the respondents**: How will you gain access to the sample you want to study? Are there gatekeepers that you need to approach to obtain permission to administer the survey?
3. **The logistics of administering the surveys**: How will you administer and score the surveys? For example, will you administer them electronically via email, face-to-face, or by phone?
4. **The possible costs involved in administering the survey**: Are there costs involved in administering the survey? Will you have to pay in order to receive a summary of responses if you use a commercial digital/electronic website to administer the survey? For example, websites such as SurveyMonkey may charge you for using their services, but they can also provide a convenient summary of the data and helpful graphs.

5. **Essential information**: What information is essential for you to obtain and what information is secondary? Clarify in your mind the most important questions you want to investigate. Just because something is "interesting to know about" does not mean you should include it in your survey. Weigh the length that such questions will add versus the information they will provide. Most people are less likely to respond to longer surveys or provide thoughtful responses when the survey takes too long to complete.

6. **Demographic information**: What identifying or demographic information is essential for you to collect? Having some demographic information from the respondents will allow you to determine whether they accurately represent the target population that you want to study. Keep in mind, though, that many respondents are reluctant to answer personal questions such as their age or annual income. If such information is essential to your survey, provide ranges of responses. For example, instead of asking for the respondents' exact age, provide age groups (such as "20–29, 30–39 . . .").

7. **Anonymity and confidentiality**: Can you ensure confidentiality for survey respondents? In some cases respondents are assured anonymity while in other cases they are asked to record some specific demographic information or even include their names.

After clarifying to yourself the answers to the questions presented above, prepare a blueprint for your survey, including the title of your survey and a brief explanation of the nature of the survey and how the results will be used. Note in your plan the type and number of items you want to use, the item format, and the order of the survey items. For example, many survey experts recommend that you place the demographic questions at the end of the survey. The following are some suggestions for writing and organizing the survey items:

1. **Length of the survey**: How many items will you have? Remember, the length can affect the response rate.

2. **Item format and response choices**: What formats will you use? The choice of responses offered should be appropriate and logical for the type of question asked or statement made. This decision will also affect the types of data analyses and statistical tests you will be able to carry out.

3. **Coding and scoring the survey items**: What numerical value will you assign to each response choice?

4. **The order of the survey items**: Are the items logically organized? Are similar items grouped together? Are all the items numbered? Are items organized from the general to the more specific?

5. **Clear instructions and language**: Are the terms used in the survey clearly defined and at the appropriate level for those responding? Are there clear instructions for

the tasks to be completed (for example, "Rank order the following choices" or "Choose the response that best reflects your opinion")?

6. **Missing data:** How will you deal with missing data? While it is always a good idea to avoid the problem of missing data, this can happen. If you have a large sample size, this is less of an issue compared with a small sample.

7. **Analyzing the data:** How will you analyze the data collected? The types of items in your survey often dictate the types of statistical tests and other analyses that you will be able to use.

8. **Pilot testing the survey:** Can you pilot test the survey by administering it to people who are like your target audience? It is always a good idea to use feedback from the pilot testing to modify the survey, as needed.

Consider consulting textbooks and the Internet on books and websites that discuss survey design and offer guidance on writing survey items. You may also be able to locate existing surveys on your topic and discover ideas about ways to conduct your survey, including suggestions about types of questions and their wording, Make sure, though, that you do not plagiarize these surveys!

ANALYZING SURVEY DATA

Analyzing data obtained via surveys can be done by novices as well as by professional statisticians. Simple tallies or frequency tables can be created and do not require any experience or expertise in data analysis. The data collected can then be used to draw conclusions. The tables can be generated by hand using simple tally marks; they can also be produced using electronic or digital survey tools. Researchers with experience and understanding of statistical analysis can further apply their knowledge to the task of survey data analysis. The important point to keep in mind is that the analysis you choose should be appropriate for the type of data collected. For example, you should not try to find the mean score for respondents' genders when you assign 1 to males and 2 to females (or vice versa) because the numbers (1 and 2) in this example comprise a nominal scale where the numerical values are assigned arbitrarily and do not imply "more than" or "less than." (See chapter 2.) The numbers used in rating-scale responses are considered interval and thus you can use statistical tests such as Pearson correlation, t test, or ANOVA to analyze the data.

In the following section, we discuss several options for analyzing survey data. The options, which are thoroughly discussed in previous chapters, are frequency tables, correlation, t test, ANOVA, and chi square. (See chapters 3, 7, 9, 10, and 11, respectively.) Of course, other statistical tests as well as graphic representations can be used to summarize and analyze survey data. The following examples are just some illustrations that will allow you to apply your newly acquired knowledge of statistical analysis.

Summarizing Survey Data Using Frequency Tables

The process of summarizing survey responses can be easily accomplished using simple tally marks as each survey is recorded. For instance, in a study of how they get to school in the morning, first grade students can add a tally mark next to choices such as walking, riding bikes, or getting a ride from parents or other adults. These tally marks can then be converted into percentages to better facilitate the interpretation of the data.

Another example is a study to discover students' future plans for postsecondary education where sixty-three students from two high schools were asked about their long-range plans and the highest degree they plan to obtain. The students had to select one of four options: (a) High School Only, (b) BA, (c) MA, and (d) Doctorate. Table 17.1 shows the responses of the students from the two schools. For easier interpretation, the recorded responses can be converted into percent (see table 17.2).[3]

Table 17.1. A Table Showing the Use of Tally Marks to Summarize Survey Responses: Postsecondary Plans of Students from Two High Schools

School	High School	BA	MA	Doctorate	TOTAL
School A	ꟼꟼꟼꟼꟼ	ꟼꟼꟼꟼꟼ ꟼꟼꟼꟼꟼ ꟼꟼꟼꟼꟼ II	ꟼꟼꟼꟼꟼ IIII	II	33
School B	III	ꟼꟼꟼꟼꟼ ꟼꟼꟼꟼꟼ ꟼꟼꟼꟼꟼ	ꟼꟼꟼꟼꟼ ꟼꟼꟼꟼꟼ I	II	31

Table 17.2. A Table Showing a Summary of the Responses in Table 17.1 Converted into Percent

School	High School	BA	MA	Doctorate	TOTAL
School A	5 (15%)	17 (52%)	9 (27%)	2 (6%)	33
School B	3 (10%)	15 (48%)	11 (35%)	2 (6%)	31
TOTAL	8 (12%)	32 (50%)	20 (32%)	4 (6%)	64

In examining the responses of the students from the two schools, it seems that the trends in the responses are fairly similar. The majority of students in both schools plan to pursue a college degree (twenty-eight from school A and twenty-eight from school B). The majority of students in both schools plan to get a BA (52 percent in school A and 48 percent in school B), followed by MA (27 percent in school A and 35 percent in school B). These data can also be presented graphically using a bar diagram with joint bars. (See chapter 3.)

Using Pearson Correlation to Analyze Survey Data

Pearson correlation can be applied to survey items that provide responses that are measured on an interval or ratio scale (for example, rating scales). For example, a study could be conducted to compare parents' attitudes toward charter schools to their opinions about the public schools in their district.

---

3. HINT: When reporting the percent of each response, remember to include the "head count" as well to communicate to your audience the actual number of responses in each category.

Correlation can also be used to compare the responses of people to survey items that comprise two or more scales. The expectation is that items that belong to the same scale would have higher correlation with each other compared with items from two different scales. This can be used as evidence that the items on a given scale indeed measure the same concept. An intercorrelation table can then be used to record and analyze these correlations. (See chapter 7.)

Another example where the survey results can be analyzed using Pearson correlation is a study conducted in a middle school district prior to implementing a new, more intense, service learning curriculum. School staff and parents design a survey to study the relationship between students' attitudes toward community service and the total number of hours that they have volunteered in the community in the last three months. The study includes the responses of thirty randomly selected students from each grade level, for a total of ninety students from grades 6–8. The survey includes five items with four response choices: Strongly Agree (4 points); Agree (3 points); Disagree (2 points); and Strongly Disagree (1 point). The school administrators and teachers want to investigate whether those with a more positive attitude toward community service also volunteer more hours in their community. Table 17.3 shows the responses of ten students in the study although the correlation coefficient that is reported was computed based on the scores of all ninety students.

Table 17.3. Sample Scores of Ten Middle School Students on Survey Measuring Attitudes toward Community Service and the Number of Hours the Students Have Volunteered in the Last Three Months

Student ID#	Score on Survey of Attitude toward Community Service	Hours per Month of Volunteering
1	7	1
2	20	6
3	6	1
4	18	5
5	15	4
6	16	5
7	20	9
8	19	7
9	6	0
10	5	1

The obtained correlation based on the scores of all ninety students in the study is $r = .81$, significant at $p < .001$. Thus, there is a strong and statistically significant relationship between students' attitudes and the number of hours that they have volunteered. As you remember, though, correlation does not imply causation; we cannot assume that attitudes are responsible for students' volunteering. The school staff and parents can implement a strong service learning program to encourage students to participate and volunteer in their community and use other statistical tests such as paired-samples *t* test to ascertain whether there is an improvement in the students' attitudes and in the number of hours that they volunteer in the community.

Using *t* Tests to Analyze Survey Data

When researchers want to compare the mean responses to items on a survey, the *t* test can be used to analyze the results. For example, a *t* test for independent samples can be used to analyze differences in survey responses between two independent groups; a *t* test for paired samples can be used to compare pretest to posttest survey means for the same group of respondents. (See chapter 9.) Following are examples of studies where we demonstrate the use of both types of *t* tests to analyze survey results.

t Test for Independent Samples

The *t* test for independent samples can be used to compare the responses of two separate groups that are responding to the same survey. For example, school staff (teachers, support staff, and administrators) can design a survey to study possible differences in students' attitudes toward cyberbullying by comparing the responses of two different groups, such as male and female students, or students who report having been bullied and those who were not bullied.

To illustrate the use of the *t* test for independent samples, let's look at another example of a study that was conducted in two states to assess teachers' opinions about the level of success in implementing the Common Core State Standards in their school. Appropriate curriculum materials have to be developed to ensure that students will meet the standards. In one state (state A) with a larger budget, teachers are using commercial materials developed by major publishers. These curriculum materials were prepared specifically to follow the Common Core curriculum for each grade level. In the other state (state B), which has more limited financial resources, the teachers have developed their own materials or are using existing materials to which they have added lesson plans, supplementary materials, and curriculum materials (including online resources). At the end of the first year, a survey is administered to teachers in the two states to gauge their level of satisfaction and attitudes toward the implementation and success of the Common Core in their schools. The 30-item survey includes several formats (for example, rating, ranking, and open-ended questions). To demonstrate the use of the *t* test for independent samples and to simplify the computations, we randomly chose the responses of fifteen teachers from each state (state A and state B) to six Likert-scale items, each with five response choices ranging from Strongly Agree (5 points) to Strongly Disagree (1 point). Thus, total scores on the six items can range from 6 to 30 (table 17.4).[4] The *t* test results are reported in table 17.5 followed by table 17.6 which reports the *t* test summary statistics.

---

4. HINT: Although we chose fifteen surveys from each state, remember that the *t* test for independent samples can be used with groups that have unequal size. Nevertheless, it is always a good idea to try to have similar sample sizes in both groups.

Table 17.4. Total Scores on Select Items from a Survey Measuring Teachers' Opinions toward the Common Core: A Comparison of the Responses of Teachers from Two States

Responses of Teachers from State A	Responses of Teachers from State B
21	19
20	24
25	28
30	30
13	28
16	22
14	21
19	19
18	27
15	22
17	18
24	22
14	9
18	10
23	19

Table 17.5. *t* Test Summary Statistics from a Survey Administered to Teachers in Two States about the Common Core

State	N	Mean	SD	t value	p value
State A	15	19.13	4.779	−1.234	.308
State B	15	21.20	6.050		

The results show that the mean attitude score of the teachers in state B was 2 points higher than the score of teachers in state A but the difference was not statistically significant. Of course, this sample size is very small and represents only a minute portion of all those who responded to the survey in both states, so conclusions cannot be drawn from these data about the efficacy of using commercially prepared curriculum materials or teacher-prepared materials.

t Test for Paired Samples

Another type of *t* test discussed in chapter 9 is aimed at comparing the means of two paired samples. The most common use of this *t* test is in studies comparing pretest scores to posttest scores obtained from the same group. For example, classroom teachers and park rangers may want to assess fourth grade students' attitudes toward and appreciation of nature and the environment as a result of an overnight field trip to a state park. A survey measuring the students' attitudes and appreciation is administered before and after the field trip. The hypothesis is that the posttest survey mean will be higher than the pretest mean.

To demonstrate the computation of the t test for paired samples that is used to analyze survey data, let's look at the next example. Fourth grade teachers at a local school participated in a day-long workshop on increasing student engagement and the relationship between engagement and student performance. The teachers then met and discussed ways to apply their new knowledge to their own classrooms. Modifications made by teachers included assigning more tasks that have clear relevance to real life, differentiating instruction, giving students more choices in the activities and assignments, communicating clear assessment goals with students' input, and providing more collaborative assignments.

To evaluate the effectiveness of the program, the teachers developed a 10-item student engagement self-assessment survey, to be completed by the students at the beginning and at the end of the school year. The survey includes items such as "I enjoy working with my team"; "I enjoy the feeling of completing my assigned work"; and "I understand the importance of setting goals and working toward achieving my goals." Response choices on the survey items range from Strongly Disagree (1 point) to Strongly Agree (4 points). Total scores on the survey can range from 10 to 40. Seventy-five students from three classrooms completed the survey. Although table 17.6 shows the results for a partial listing of the data, the t test analysis included all seventy-five students. The results of the paired-samples t test are presented in table 17.7.

Table 17.6. Responses of Ten Fourth Grade Students to a Survey Measuring Students' Level of Engagement Self-Assessment

Student	Beginning of the Year Survey Score	End of the Year Survey Score
1	24	30
2	31	40
3	15	32
4	26	35
5	12	28
6	35	40
7	18	33
8	23	28
9	11	31
10	17	36

The results displayed in table 17.7 show a significant increase in student self-engagement. The mean at the beginning of the year was 16.73 and 29.16 at the end of the year. The t value was fairly high ($t = 16.600$) and it was statistically significant ($p < .001$). The program seems to be very effective in increasing students' self-assessment of their level of engagement.

Table 17.7. Results of Paired-Samples t Test Comparing Pretest and Posttest Engagement Self-Assessment Survey Scores of Seventy-Five Fourth Grade Students

	N	Mean	SD	t value	p value
Pretest	75	16.73	6.550	−16.600	<.001
Posttest	75	29.16	7.903		

One-Way ANOVA

One-way analysis of variance (ANOVA) can be used to analyze survey data measured on an interval or ratio scale, when gathered from two or more independent groups. For example, researchers may want to evaluate college students' attitudes toward drinking on campus by administering a survey and comparing the responses of freshmen, sophomores, juniors, and seniors.

To show the application of one-way ANOVA to survey data, let's look at the following example. School board members and administrators of a K–12 school district wanted to gather information about parents' opinions toward school. They prepared a 10-item survey with possible scores ranging from 10 to 50. The results were analyzed separately for parents of students in grades K–5, 6–8, and 9–12. Items on the survey included: "I am pleased with my child's teachers" and "The amount of homework is appropriate to my child's grade level." To demonstrate the computations and interpretation of the data, we selected at random the scores of twenty-five parents from each group (K–5, 6–8, and 9–12). Table 17.8 shows the groups' means and standard deviations, as well as the F ratio and p values.

Table 17.8. A Survey Summary Table Showing the Means, Standard Deviation, F Ratio, and p Value of Grades K–5, 6–8, and 9–12

Grade Level	N	Mean	Standard Deviation	F ratio	p value
K–5	25	35.24	6.547	1.243	.295
6–8	25	33.64	7.111		
9–12	25	32.08	7.560		
TOTAL	75	33.65	7.108		

As can be seen from the data in the table, the K–5 parents have the most positive attitude (the highest mean), followed closely by the parents of students in grades 6–8 and 9–12. However, the differences between the three groups of parents are small and not statistically significant ($p = .295$). The school board members and administrators may benefit from an item analysis where they can determine which items on the survey elicited the most positive responses and which items generated lower scores, indicating that some improvements are needed.

Using the Chi Square Test of Independence to Analyze Survey Data

The chi square test can be used to analyze survey responses that are measured on a nominal scale (see chapter 2). For example, anticipating a sharp increase in school-age children in the district in the next few years, a study is carried out by the school board to gauge the support of people living in the district for building a new elementary school. The school board sends a brief survey that includes two statements asking respondents to choose one of them: "I support building a new school in the district" and "I oppose building a new school in the district." Data can be analyzed by comparing the responses of parents with school-age children in public school to those of other district residents without school-age children or those with children in private schools. The results can be recorded in a 2 × 2 table, where the rows represent the two groups of respondents, and the columns represent the response choices ("I support building a new school" or "I oppose building a new school").

Let's look at another study using the chi square test of independence to analyze survey data. In this inquiry, school leaders of a K–5 school want to evaluate students' attitudes toward a revamped lunch menu that is intended to be more nutritious as well as tasty and appealing. Menu items such as vegetarian burgers on whole wheat buns, oven-roasted potatoes, and carrot sticks are added for lunch. The responses of the primary-grades students (K–3) are compared to those of the upper-grades students (4–5). Response options include: (a) I like it; (b) I do not like it; and (c) I have no opinion. The responses of 198 students at the school who have tried the new lunch menu are summarized in table 17.9.

Table 17.9. A Comparison of the Levels of Support for the New Lunch Menu of Students in Grades K–3 and Grades 4–5

Grades	I Like It	I Don't Like It	I Have No Opinion	TOTAL
K–2	66	21	10	97
4–5	55	39	7	101
TOTAL	121	60	17	198

The chi square value is $\chi^2 = 6.51$, significant at $p = .033$. As is evident by the results displayed in table 17.7, while the majority of students in both grade levels like the new menu selection, the preference is more evident in the primary grades and the difference between the two grade levels is statistically significant.

SUMMARY

1. A **survey** is usually undertaken to gather information on a select group of respondents about a topic that is of interest to the researchers.
2. Information obtained using surveys may include factual information or information about respondents' behavior, attitudes, and opinions.

3. Several item formats are briefly discussed; the data they yield can be measured on nominal, ordinal, interval, or ratio scales and therefore may call for different statistical analyses.
4. Although surveys may include open-ended items that call for narrative responses, this chapter focuses on numerical data only.
5. The chapter discusses various ways to plan, design, create, administer, and analyze surveys.
6. Examples of how to use frequency tables to record survey data are presented.
7. To use Pearson correlation, independent-samples t test, paired-samples t test, and one-way analysis of variance (ANOVA), data measured on interval or ratio scales must be used.
8. When the response choices are presented as nominal, discrete options, the chi square test may be used to analyze the data.

CHECK YOUR UNDERSTANDING

1. Design a survey on a topic of your choice:
   a. Explain what types of items you will include and the rationale for these items.
   b. Describe how you will pilot test and administer the survey.
   c. Propose and explain the statistical tests you will undertake to analyze your data.
2. Describe a survey where the data collected can be analyzed using Pearson correlation.
3. Describe a study where the data collected can be analyzed using t test or ANOVA.
4. Describe a study where the data collected can be analyzed using a chi square test.

Glossary

A-B-A single-case design: Single-case experimental designs with three phases: A (baseline); B (intervention); and A (a second baseline, after the intervention is withdrawn). Multiple data points are used at each phase to obtain a stable measure of the target behavior. (Ch. 1)

Action research: Also *practitioner research*; research that is undertaken by practitioner-researchers to study their own setting. Action research is usually conducted by practitioners to solve a problem by studying it, proposing solutions, implementing the solutions, and assessing the effectiveness of these solutions. Both qualitative and quantitative data can be collected in action research. (Ch. 1)

Alternate forms reliability: An approach used to assess the degree of consistency between two forms of the same test. (Ch. 13)

Alternative hypothesis: A prediction about the expected outcomes of the study that guides the investigation and the design of the study. The alternative hypothesis is represented by H_A or H_1. Often, the alternative hypothesis is simply referred to as *the hypothesis*. It often predicts that there would be some relationship between variables or a difference between groups or means. (Ch. 2)

Amodal distribution: A distribution without a mode. (Ch. 4)

Analysis of variance (ANOVA): A statistical test used to compare the means of two or more independent samples using interval or ratio scale data, and to test whether the differences between the means are statistically significant. (Ch. 10)

Applied research: Research that is aimed at testing theories and applying them to specific situations. Based on previously developed theories, hypotheses are developed and tested in studies classified as applied research. (Ch. 1)

Bar graph (or bar diagram): A graph with a series of bars that do not touch that is used to display *discrete* and independent categories or groups. The bars are often ordered in some way (for example, from highest to lowest). (Ch. 3)

Basic research: Research that is conducted mostly in labs, under tightly controlled conditions, and its main goal is to develop theories and generalities. This type of research is not aimed at solving immediate problems or at testing hypotheses. (Ch. 1)

Bimodal distribution: A distribution with two modes. (Ch. 4)

Box plot (or box-and-whiskers): A graph that is used to show the median and spread of a set of scores using a box and vertical lines. The two middle quartiles are located *within* the box, and a horizontal line inside the box shows the location of the median. The two extreme quartiles are displayed using the vertical lines (the "whiskers") *outside* the box. (Ch. 3)

Causal comparative (or *ex post facto*) research: Research designed to study cause-and-effect relationships, where the independent variable is not manipulated because it occurred prior to the start of the study or it is a variable that cannot be manipulated. (Ch. 1)

Chi square (χ^2) test: A nonparametric statistical test that is applied to categorical or nominal data where the units of measurement are *frequency* counts. *Observed* frequencies gathered in a study are compared to *expected* frequencies to test whether the differences between them are significant. The chi square test statistic is represented by χ^2. There are two types of chi square tests: (a) the *goodness of fit* chi square test that is used with one independent variable; and (b) the chi square *test of independence* that is used with two independent variables. (Ch. 11)

Class intervals: Equal-width groups of scores in a distribution. (Ch. 3)

Coefficient alpha: See Cronbach's coefficient alpha (Ch. 13)

Coefficient of determination (r^2): An index used to describe the proportion of variance in one variable (usually the criterion) that can be explained by differences in the other variable (usually the predictor). The coefficient of determination may also be called the *shared variance*. (Ch. 7)

Cohort study: A type of longitudinal study where similar people, selected from the same cohort, are studied at two or more points in time. (Ch. 1)

Concurrent validity: The correlation between scores from an instrument and scores from another well-established instrument that measure the same thing. (Ch. 14)

Confidence interval (CI): A range within which we would expect to find, with a certain level of confidence (for example, 95 percent), the population value we want to estimate from our sample. The interval includes two boundaries: a lower limit (CI_L) and an upper limit (CI_U). (Ch. 2)

Constant: A measure that has only one value (Ch. 2). In *regression*, the constant is the point where the regression line intersects the vertical axis. (Ch. 8)

Construct validity: The extent to which a test measures and provides accurate information about a theoretical trait or characteristic. (Ch. 14)

Content validity: The degree to which an instrument measures a representative sample of behaviors and content domain about which inferences are to be made; the extent of the match between the test and the content it is intended to measure. (Ch. 14)

Continuous variable: A variable that can take on a wide range of values and contain an infinite number of increments. (Ch. 2)

Convenience (or incidental) sample: A sample that is chosen for the study by the researcher because of its convenience. (Ch. 2)

Correlation: The relationship or association between two or more paired variables. (Ch. 7)

Correlation coefficient: An index indicating the degree of association or relationship between two variables. The coefficient can range from –1.00 (perfect negative) to +1.00 (perfect positive): 0.00 indicates no correlation. The most commonly used coefficient is the Pearson's *r*. (Ch. 7)

Counterbalanced designs: Experimental designs where several interventions are tested simultaneously, and the number of groups in the study equals the number of interventions. All the groups in the study receive all interventions, but in a different order. (Ch. 1)

Criterion-referenced (CR) test: A test used to compare the performance of an individual to certain criteria. (Ch. 12)

Criterion-related validity: The degree to which an instrument is related to another measure, called the criterion. (See **concurrent validity** and **predictive validity** for types of criterion-related validity.) (Ch. 14)

Critical value: A cutoff point between statistical results that are considered *statistically significant* and those that are *not statistically significant*. The computed test statistics are compared to the appropriate critical values in order to make decisions about whether to retain or reject the null hypothesis. (Ch. 2)

Cronbach's coefficient alpha: A type of internal consistency measure of test reliability that uses scores from a single test administration. This approach assesses how well items or variables that measure a similar trait or concept correlate with each other. (Ch. 13)

Cross-sectional designs: Nonexperimental designs conducted to study how individuals change and develop over time by collecting data at one point in time on different-age individuals. (Ch. 1)

Cumulative frequency distribution: A distribution of scores that shows the number and percentage of scores *at* or *below* a given score. The distribution includes the following: scores, frequencies, percent frequencies, cumulative frequencies, and cumulative percent frequencies. (Ch. 3)

Degrees of freedom (*df*): In most cases, the degrees of freedom are $n - 1$ (the number of people in the study, minus 1), although there are some modifications to this rule in some statistical tests. (Ch. 2)

Dependent variable: An outcome measure in an experimental study designed to measure the effectiveness of the intervention. (Ch. 1) The criterion variable in regression. (Ch. 8)

Descriptive research: Studies aimed at investigating phenomena as they are naturally occurring, without any manipulation or intervention. (Ch. 1)

Descriptive statistics: Procedures used to classify, organize, and summarize numerical data about a particular group of observations. There is no attempt to generalize these statistics, which describe only one group, to other samples or populations. (Ch. 2)

Deviation score: The distance between each score in a distribution and the mean of that distribution, expressed as: $x-\bar{x}$. (Ch. 5)

Differential selection: A threat to internal validity; refers to studies where preexisting group differences may contribute to different performances of members of these groups on the dependent variable. (Ch. 1)

Directional hypothesis: A prediction that states the *direction* of the outcome of the study. In studies where group differences are investigated, a directional hypothesis predicts which group's mean would be higher; and in studies that investigate relationships between variables, a directional hypothesis predicts whether the correlation will be positive or negative. (Ch. 2)

Discrete variable: A variable that contains a finite number of distinct values between any two given points. (Ch. 2)

Effect size (ES): An index that is used to express the strength or magnitude of difference between two means or the strength of association of two variables. The comparison of the means is done by converting the difference between the means into standard deviation units. Effect size can also be used to assess the strength of the association between two variables by using the correlation coefficient (r) or a square of the correlation coefficient (r^2, or R^2). (Ch. 2)

Experimental research: Research designed to study cause-and-effect relationships by manipulating the *independent* variable (that is, the cause) and observing possible changes in the *dependent* variable (the effect, or outcome). Experimental research is designed to assess the effectiveness of a planned intervention on groups or individuals. (Ch. 1)

***Ex post facto* research:** See **Causal comparative research**. (Ch. 1)

External validity: The extent to which the results of the study can be generalized and applied to other settings, populations, and groups. (Ch. 1)

Extraneous variable: A variable that presents a threat to the study's internal validity; an uncontrolled variable that can present a competing explanation for the impact of the planned intervention. (Ch. 1)

F **ratio (or** *F* **value):** A test statistic used in analysis of variance (ANOVA). It is computed by dividing two variance estimates by each other. (Ch. 10)

Face validity: The extent to which an instrument *appears* to measure what it is intended to measure. (Ch. 14)

Factorial ANOVA: A general name for ANOVA with two or more independent variables. (Ch. 10)

Frequency distribution: A distribution of scores that are ordered and tallied. (Ch. 3)

Frequency polygon: A graph that is used to display frequency distributions. The bell-shaped normal distribution is a special case of a frequency polygon with a large number of cases. (Ch. 3)

Grade equivalent (GE): A scale that is used to convert raw scores to grade-level norms by expressing scores in terms of years and months. (Ch. 12)

Graph: A visual representation of numerical data. (Ch. 3)

Hawthorne Effect: A threat to external validity whereby the behavior of the study's participants may be affected by their knowledge that they are participating in a study, rather than by the planned intervention. (Ch. 1)

Histogram: A graph that contains a series of consecutive vertical bars used to display frequency distributions. (Ch. 3)

History: A threat to internal validity; refers to events that happened during the study that may affect the dependent variable. (Ch. 1)

Hypothesis: A prediction about the outcome of the study; an "educated guess." (Ch. 2)

Independent variable: The intervention (or treatment) in experimental studies; the grouping variable in nonexperimental studies. (Ch. 1) The predictor variable in regression. (Ch. 8)

Inferential statistics: Procedures that involve selecting a sample from a defined population and studying that sample in order to draw conclusions and make inferences about the population. The sample that is selected is used to obtain sample statistics to estimate the population parameters. May also be called *sampling statistics*. (Ch. 2)

Instrumentation: A threat to internal validity; refers to the level of reliability and validity of the instrument being used to assess the effectiveness of the intervention. (Ch. 1)

Interaction: A situation in factorial ANOVA where one or more levels of the independent variable, when combined with another independent variable, have a different effect on the dependent variable. (Ch. 10)

Intercorrelation table: A table that is used to display the correlations of several variables with each other. (Ch. 7)

Internal consistency approach: Approaches used to assess the reliability of an instrument using scores from a single administration of the instrument. (Ch. 13)

Internal validity: The extent to which observed changes in the dependent variable (outcome measure) can be attributed to the independent variables (the intervention); the extent of control over the extraneous variables. (Ch. 1)

Inter-rater reliability: A method to assess the degree of consistency and agreement between scores assigned by two or more raters or observers who judge or grade the same performance or behavior. (Ch. 13)

Interval scale: A measurement scale with observations that are ordered by magnitude or size with equal intervals between the different points. (Ch. 2)

John Henry Effect: A threat to external validity; refers to a condition where the intervention does not seem to be effective because control group members perceive themselves to be in competition with experimental group members and consequently perform above and beyond their usual level. (Ch. 1)

Level of significance (*p* level): The level of error associated with rejecting a null hypothesis; the probability that the study's results were obtained purely by chance. (Ch. 2)

Linear regression: A process of prediction where the predictor variable (X) and the criterion variable (Y) have a linear relationship. (Ch. 8)

Line graph: A graph used to show relationships between two variables through lines that connect the data points. The *horizontal* axis indicates values that are on a continuum, and the *vertical* axis can be used for various types of data. (Ch. 3)

Longitudinal studies: Nonexperimental designs conducted to measure changes over time by following the same group of individuals. (Ch. 1)

Maturation: A threat to internal validity; refers to physical or mental changes experienced by the study's participants while the study takes place. (Ch. 1)

Mean: The most commonly used measure of central tendency that is obtained by adding up the scores and dividing the sum by the number of scores; also called the *arithmetic mean.* (Ch. 4)

Mean squares (MS): In ANOVA, there are different variance estimates. For example, MS_W is the estimate of the variances *within* groups; MS_B is the estimate of the variance of groups around the total mean. (Ch. 10)

Measure of central tendency: A summary score; a single score that represents a set of scores. (Ch. 4)

Measurement: A process of assigning numbers to observations according to certain rules. (Ch. 2)

Median: A measure of central tendency that is the distribution's midpoint, where 50 percent of the scores are above it and 50 percent are below it. (Ch. 4)

Mode: A measure of central tendency; the score that occurs with the greatest frequency. (Ch. 4)

Multimodal distribution: A distribution with three or more modes. (Ch. 4)

Multiple correlation (*R*): An index of the combined correlation of the predictor variables with the criterion variable. (Ch. 8)

Multiple regression: See **Regression**.

Nominal scale: A measurement scale where numbers are used to label, classify, or categorize data. The various points on the scale are not ordered and the distances between them are not equal. (Ch. 2)

Nondirectional hypothesis: A hypothesis that predicts that there would be a difference or relationship, but the direction of the difference or association is not specified. (Ch. 2)

Nonexperimental research: A research study where no planned intervention takes place. Nonexperimental research is divided into two types: *causal comparative* (also called *ex post facto*) and *descriptive*. (Ch. 1)

Nonparametric statistics: Statistics that are used with ordinal and nominal data or with interval and ratio scale data that fail to meet the assumptions needed for parametric statistics. Nonparametric statistics are easier to compute and understand, compared with parametric statistics. (Ch. 2)

Normal curve: A graphic presentation of a theoretical model that is bell-shaped. Various characteristics in nature are normally distributed, and each normal distribution has its own mean and standard deviation. (Ch. 6)

Normal distribution: A symmetrical distribution where the mean, median, and mode have the same value and the scores tend to cluster in the center; about two-thirds of the scores are within $\pm 1SD$ from the mean. (Ch. 6)

Norming group: A group used to develop test norms with demographic characteristics similar to those of the potential test-takers. (Ch. 12)

Norm-referenced (NR) test: A test that includes norms designed to compare the performance of examinees taking the test to the performance of similar individuals in a norming group who took the same test and whose scores were used to generate the norms. (Ch. 12)

Null hypothesis: A hypothesis that predicts that there would be no relationship between variables or no difference between groups or means beyond that which may be attributed to chance alone; represented by H_O. In most cases, the null hypothesis (which may also be called the *statistical hypothesis*) is not formally stated, but it is always implied. (Ch. 2)

One-tailed test: Used when the alternative hypothesis (that is, the study's main research hypothesis) is directional, to decide whether to reject the null hypothesis. (Ch. 7, 9)

One-way analysis of variance (one-way ANOVA): Analysis of variance with one independent variable and one dependent variable. (Ch. 10)

Ordinal scale: A measurement scale where the observations can be ordered based on their magnitude or size and the intervals among the different points on the scale are *not* assumed to be equal. (Ch. 2)

Outlier: A score that is noticeably different from the other scores in the distribution and is outside the range and pattern of the other points; an extreme score. (Ch. 7)

p (probability) level (level of significance): An index used to indicate the *probability* that we are making an error in rejecting a true null hypothesis. A probability level of 5 percent is commonly used to decide whether to consider the results statistically significant. (Ch. 2)

Panel study: A type of longitudinal study where the same people are studied at two or more points in time. (Ch. 1)

Parameter: A measure that describes a characteristic or value of an entire population. (Ch. 2)

Parametric statistics: Statistics that are applied to data from populations that meet the following assumptions: the variables being studied are measured on an interval or a ratio scale; cases (for example, participants) are randomly assigned to groups; the scores are normally distributed; and the variances of the groups being compared are similar. Parametric tests are considered more efficient and powerful than their nonparametric counterparts. (Ch. 2)

Pearson's r: A measure of the linear relationship between two continuous variables measured on an interval or ratio scale. Pearson's r can range from -1.00 (a perfect negative correlation) to 0.00 (no correlation) to $+1.00$ (perfect positive). (Ch. 7)

Percentile band: An estimated range where a student's true percentile rank is expected to be, usually reported with 68 percent confidence level. (Ch. 12)

Percentile rank: An index that describes the relative position of a score obtained by a person on a scale by indicating the percentage of people at or below that score. (Ch. 3, 6, 12)

Pie graph (or pie chart): A graph that looks like a circle divided into "wedges" or "segments." Each wedge represents a category or subgroup within that distribution. (Ch. 3)

Population: An entire group of persons or elements that have at least one characteristic in common. (Ch. 2)

Post hoc comparison: In ANOVA, it is a process of *multiple comparisons* done *after* the completion of the study where all possible pairs of means are compared to determine which differences between the means are statistically significant. *Tukey's honestly significant difference (HSD)* is an example of a post hoc comparison test. (Ch. 10)

Practitioner research: See **Action research.** (Ch. 1)

Predictive validity: The extent to which an instrument can predict some future performance. (Ch. 14)

Preexperimental designs: Designs classified as preexperimental do not have a tight control over extraneous variables, and their internal validity cannot be assured. (Ch. 1)

Qualitative research: Research that seeks to understand social or educational phenomena. The researcher focuses on one or a few cases that are studied in depth using multiple data sources that are subjective in nature. Qualitative research, which uses mostly narrative data, is context-based, recognizing the uniqueness of each individual and setting. (Ch. 1)

Quantitative research: Research that is conducted to describe phenomena or to study cause-and-effect relationships by examining a small number of variables and using numerical data gathered from large samples. Researchers conducting quantitative research usually maintain objectivity and detach themselves from the study's environment. (Ch. 1)

Quasi-experimental designs: Experimental designs where intact groups are used and where the groups being compared are not assumed to be equivalent at the beginning of the study. (Ch. 1)

Range: A measure of spread (or variability) that indicates the distance between the highest and lowest scores in a distribution. (Ch. 5)

Ratio scale: A measurement scale where the observations are ordered by magnitude, with equal intervals between the different points on the scale and an absolute zero. (Ch. 2)

Raw score: A score obtained by an individual on some measure that is not converted to another measure or scale. (Ch. 4)

Regression: A statistical technique used for estimating scores on one variable (the dependent variable, or criterion) from scores on one (or more) variable (the independent variable, or predictor). When one variable is used to predict another, the procedure is called *simple regression*, and when two or more variables are used as predictors, the procedure is called *multiple regression*. (Ch. 8)

Regression line: A line of best fit on a scattergram where the predicted scores are expected to be. (Ch. 9)

Reliability: The level of consistency of an instrument and the degree to which the same results are obtained when the instrument is used repeatedly with the same individuals or groups. (Ch. 13)

Research: A systematic inquiry that includes data collection and analysis. The goal of research is to describe, explain, or predict present or future phenomena. There are several ways to classify research, and each approach looks at research from a different perspective. (Ch. 1)

Sample: A small group of observations selected from the total population for the purpose of making inferences about the population. A sample should be *representative* of the population because information gained from the sample is used to estimate and predict the population characteristics that are of interest. (Ch. 2)

Sample bias: *Systematic,* rather than *random,* differences between the population and the selected sample; a systematic error in a sample. (Ch. 2)

Sampling error: A *chance* variation in the numerical values of a sample (for example, mean) that occurs when we repeatedly select same-size samples from the same population and compare their numerical values. Sampling error is beyond the control of the researcher. (Ch. 2)

Scattergram (or scatterplot): A graph used to depict the association (correlation) between two numerical variables. (Ch. 7)

Simple random sample: A sample where every member of the population has an equal and independent chance of being selected for inclusion. (Ch. 2)

Simple regression: See **Regression.** (Ch. 8)

Single-case (or single-subject) designs: Experimental designs where individuals are used as their own control. Their behavior or performance is assessed during two or more phases, alternating between phases *with* and *without* an intervention. See also **A-B-A single-case designs.** (Ch. 1)

Split-half method: A procedure for assessing test reliability by dividing the items into two halves and correlating the scores from one half with the other. *Spearman-Brown prophecy formula* is then used to estimate the reliability of a full-length test. (Ch. 13)

Standard deviation (SD): A measure of spread in a distribution of scores; the mean of the distances of the scores around the distribution mean. The standard deviation is the squared root of the variance. The SD of the sample is S, and the SD of the population is σ (the Greek letter *sigma,* lowercase). (Ch. 5)

Standard error of estimate (S_E): An index that estimates the amount of error expected in predicting a criterion score; the standard deviation of the differences between actual and predicted scores in regression. (Ch. 8)

Standard error of measurement (SEM): An estimate of the error in a person's reported test score. (Ch. 13)

Standard error of the mean: The standard deviation of the sample means, expressed by the **symbol** $SE_{\bar{x}}$. (Ch. 2)

Standard score: A derived scale score that expresses the distance of the original score from the mean in standard deviation units. The most common standard score is the z score. (Ch. 6)

Stanine: A 9-point scale that is derived from the words "**sta**ndard **nine**" with a mean of 5 and a standard deviation of 2. Stanines allow the conversion of percentile ranks into larger units. (Ch. 12)

Statistic: A measure that describes a numerical characteristic of a sample. (Ch. 2)

Statistical regression: A threat to internal validity; refers to a phenomenon whereby people who obtain extreme scores on a pretest tend to score closer to the mean of their group upon subsequent testing, even when no intervention is involved. (Ch. 1)

Statistically significant: Most researchers use the convention whereby they report their findings as statistically significant if their computed probability level (p value) is 5 percent or less ($p \leq .05$). Reporting results as statistically significant means that the likelihood of obtaining these results purely by chance is low and that similar results would be obtained if the study were repeated. (Ch. 2)

Stratified sample: A sample that contains proportional representations of the population subgroups. To obtain a stratified sample, the population is first divided into subgroups (strata), then a random sample is obtained from each subgroup. (Ch. 2)

Sum of squares (SS): In ANOVA, these are different sources of variability. The *within-groups sum of squares (SS_W)* is the variability *within* the groups. The *between-groups sum of squares (SS_B)* is the average variability of the means of the groups *around* the total mean. (SS_B may also be called *among-groups sum of squares*; abbreviated as SS_A.) The *total sum of squares (SS_T)* is the variability of *all* the scores around the total mean. (Ch. 10)

Survey: A study that is usually undertaken to gather information on a selected group of respondents about a topic that is of interest to the researchers. (Ch. 17)

Systematic sample: A sample where every *Kth* member (for example, every fifth person) is selected from a list of all population members. (Ch. 2)

T score: A type of a standard score measured on a scale with a mean of 50 and a SD of 10. All the scores on the T score scale are positive and range from 10 to 90. T scores can be converted from z scores using the formula: $T = 10(z) + 50$. (Ch. 6)

t test: A statistical test used to compare two means. The means may be from two different samples, from paired samples, or from a sample and a population. The scores used to compute the means should be measured on an interval or ratio scale and be derived from the same measure. (See **t test for independent samples**, **t test for paired samples**, and **t test for a single sample**.) (Ch. 9)

t test for paired samples: A t test that is used to compare the mean scores of two sets of scores that are paired. (May also be called a t test for *dependent, matched*, or *correlated* samples.) (Ch. 9)

t test for a single sample: A t test used to compare the mean of a sample ($\bar{x}$) to the mean of a population (μ). (Ch. 9)

t test for independent samples: A t test used to compare the mean scores of two groups that are independent of each other. (Ch. 9)

Testing: When used in the context of *threats to internal validity*, testing refers to the potential effect that a pretest may have on the performance of people on the posttest. (Ch. 1)

Test-retest reliability: A procedure for assessing the reliability of a test by administering the test twice to the same group of examinees and correlating the two sets of test scores. (Ch. 13)

Time-series designs: Designs that are classified as quasi-experimental, where intact groups are tested repeatedly *before* and *after* the intervention. (Ch. 1)

Trend study: A type of longitudinal study where the same research questions are posed at two or more points in time to similar individuals. (Ch. 1)

True experimental designs: Experimental designs where the groups are considered equal because participants are randomly assigned. (Ch. 1)

Two-tailed test: Used when the alternative hypothesis (that is, the study's main research hypothesis) is nondirectional or stated as no difference between means or no association between variables. (Ch. 7, 9)

Two-way analysis of variance (two-way ANOVA): An ANOVA test used to compare two independent variables (or factors) *simultaneously*. This statistical test enables researchers to study the effect of each of the two factors on the dependent variable as well as the interaction of the two factors. The independent variables in factorial ANOVA are also called the *main effects*. (Ch. 10)

Type I error: An error made by researchers when they decide to *reject* the null hypothesis when in fact it is true and *should not be rejected*. (Ch. 2)

Type II error: An error made by researchers where they decide to *retain* the null hypothesis, when in fact it *should be rejected*. (Ch. 2)

Validity: The degree to which an instrument measures what it is supposed to measure and the appropriateness of specific inferences and interpretations made using the test scores. (Ch. 14)

Variable: A measured characteristic that can assume different values or levels. (Ch. 2)

Variance: A measure of spread in a distribution of scores, in squared units. It is the mean of the squared distances of the scores around the distribution mean. The variance can be obtained by squaring the standard deviation. The variance of the sample is S^2 and the variance of the population is σ^2 (the Greek letter *sigma*, lowercase, squared). (Ch. 5)

z score: A type of standard score that indicates how many standard deviation units a given score is *above* or *below* the mean for that group. The z scores create a scale with a mean of 0 and a standard deviation of 1. (Ch. 6)

Index

A-B-A single-case design, 14, 19
abstract in research report, 226, 233, 237
academic self-concept (ASC), 120–21
action research, 3–6, 17–18, 134
ACT test, 137–*38,* 140–41, 185, 187, 207
affective domain, 194, 201, 203
alpha level, setting, 34, 44
alternate forms reliability, 193, 195–97, 202
alternative hypothesis, 31–33, 40–41, 43–44;
    for correlation, 110–11, 117; for one-way
    ANOVA, 156, 158, 168; for *t* test, 134–35,
    138. *See also* research hypothesis
American Educational Research Association
    (AERA), 228n3
American Psychological Association (APA),
    228n3. *See also APA Publication Manual*
amodal distribution, 74, 253
among-groups sum of squares (*SSA*), 152,
    263
analysis of variance (ANOVA), 149–69;
    alternative hypothesis in, 156, 158, 168;
    a priori comparison in, 156, 177–*78,* 182;
    degree of freedom in, 153, 156, 159, 161,
    164, 166, 168–69; dependent variables
    in, 150–52, 158, 160–63, 167–69; *F* ratio
    in, 149–51, 154–59, 164, 167–68, 249;
    F statistic in, 150, 168; homogeneity
    of variances in, 150, 167–68; honestly
    significant difference in, 159; independent

variables in, 149–52, 158, 160–63, 167–69;
    pair-wise comparison in, 149, 151n; Type
    I errors in, 150. *See also* one-way analysis
    of variance (one-way ANOVA); two-way
    analysis of variance (two-way ANOVA)
analyzing survey data, 239–51
ANOVA summary table, 156–*57, 159,*
    164–*66*
APA Publication Manual, 57, 227, 232, 233n,
    234, 236
appendix in research proposal and report,
    226, 232–34
applied research, 4–5, 17
a priori comparison in ANOVA, 156, 177–
    78, 182
arithmetic mean, 72, 74
assessment of validity, 104, 205, 209
assumption-free statistics, 43
assumption of the homogeneity of variances,
    137, 146, 150, 167–68
averages, 72n. *See also* measures of central
    tendency

bar diagrams. *See* bar graphs
bar graphs, 49, 58–*61,* 66, 234
baseline measure, 14–15
basic research, 4, 6, 17
behaviorism, theory of, 4
benchmarks approach, 111

between-groups mean square (MSB), 161

between-groups sum of squares groups
(SSB), 152, 168

between groups variation, 169

bias: sample, 28–29, 43; test, 205, 209–10

bimodal distribution, 70–*71, 74*

box-and-whiskers. *See* box plots

box plots, 49, 62–*64,* 66–67

Campbell, D. T., 9n

case study approach, 6, 15, 18

causal comparative *(ex post facto)* research, 3,
7, 15, 18, 136

cause-and-effect relationships, 5, 7, 15, 18

census survey, 16

central tendency, measures of. *See* measures
of central tendency

charts. *See* graphs

chi-square goodness of fit test, 175–76, 182

chi-square ($\chi 2$ test), 173–82, 218; to analyze
survey data, 250; assumptions for, 175–
76; equal expected frequencies, 176–77,
182; expected frequencies for, 173–82;
goodness of fit test, 175–76, 182; observed
frequencies for, 173–77, 179–82; unequal
expected frequencies, 177–*78,* 182

chi-square ($\chi 2$) test of independence, 173,
175–76, 178–79, 182, 218, *220,* 250

chi square value ($\chi 2$), 174–76, 178–81, 250

class intervals, 51, 54–*55,* 60, 66–67

coefficient: alpha (Cronbach's alpha), 196,
198, 202; of determination *(r* 2), 113–14,
117–18, 126; validity, 207

cognitive domain, 194, 201, 203

Cohen, J., 35, 37, 45

cohort studies, 16–17, 20

commercial achievement tests, 206

computer statistical packages, 111, 180

concurrent validity, 207–8, 210

confidence intervals, 39–40, 46

constant, 22, 42, 96, 121, 123, 126, 128–29

construct validity, 206, 208–10

content-referenced tests, 190n

content validity, 206–7, 209–10

contingency tables, 179

continuous variables, 22, 42, 116n, 175

control group, 7–9, 11–13, 18, 32, 35–36, 39,
134–40, 146, 158, 201, 234

convenience samples, 27–28, 43

correlation, 103–18; coefficient of
determination, 113–14, 117–18,
126; curvilinear relationship, 108–9;
defined, 104; in descriptive research,
16; direction of, 104; factors affecting,
112–13; graphing, 104–8; hypotheses for,
110–11; intercorrelation table in, 115–16;
linear relationship, 108–9, 117, 120n;
magnitude of, 104; multiple, 126, 129;
negative, 37, 103–6, 108–9, 117; Pearson
product moment, 108–12, 117, 174, 194;
positive, 31, 37, 104–6, 108–9, 117, 207–8;
scattergrams in, 104–8, 117; strength
or degree of, 104. *See also* Pearson
correlation

correlation coefficient, 33, 35–37, 45, 103–4,
109–18 *passim,* 193–98 *passim,* 202, 207,
210, 216, 245; interpreting, 109–10, 113;
Pearson, 41, 103

correlation tables, 103, 116, 118

counterbalanced designs, 13, 19

criterion-referenced (CR) tests, 85, 185–86,
190, 192

criterion-related validity, 205–7, 209–10

criterion variables, 120, 122–23, 128

critical value, 33–34; chi square and, 176–78,
180–81; correlation and, 111–12; degrees
of freedom and, 34; *t* tests, 135, 139–40,
143, *145–46*

Cronbach, Lee, 198

Cronbach's coefficient alpha, 196, 198, 202

cross-sectional design, 16–17, 19–20

cumulative frequency distributions, 52, 66

curvilinear relationships, 108–9

data: analysis of, 41, 239–51; graphing,
49–50, 53–67; nominal, 36; ordinal, 29;

organizing, 49–53; qualitative, 4, 240;
quantitative, 4, 15, 226, 240. *See also*
numerical data
data analysis section in research proposal,
229, 231–32, 237
decision flowchart, choosing statistical test
for, 215–*17*
decision making, 32–33, 194, 198, 201, 203,
206; errors in, 34
degrees of freedom, 34, 44; in chi square test,
175–76, 178–81; in correlation, 111–12; *t*
test and, 140, 145; variance and, 153, 156,
159, 161, 164, 166, 168–69
dependent variables, 18–19, 44, 120,
123, 128, 175; in ANOVA, 150–52,
158, 160–63, 167–69; in experimental
research, 6, 8–10, 12, 14–15, 18, 32; in
nonexperimental research, 7, 15
descriptive research, 3, 6n, 7, 16–19;
correlation in, 16
descriptive statistics, 16, 29–30, 43;
graphing data in, 53–67; measures of
central tendency, 70, 73–74; measures
of variability, 77–87; organizing data in,
49–53; in research proposal, 232
descriptive studies, 15, 18, 232
deviation scores, 79–81, 86
differential selection as threat to internal
validity, 11, 19
directional hypothesis, 31–32, 44, 134, 138,
146
discrete variables, 22, 42
discussion in research report, 9, 226–27, 230,
233, 235, 237
disordinal interaction, 162, 169
distribution-free statistics, 29n, 181
distributions: with extreme scores, variance
and standard deviation, 84–86; skewed,
89, 93–94, 99
domain-referenced tests, 190n

educational research, 1–20; action
research in, 4–5; applied research in,

4; basic research in, 4; experimental
vs. nonexperimental research in, 6–18;
practitioner research in, 5; quantitative vs.
qualitative research in, 5–6
effect size (ES), 34–37, 44–45, 146; coefficient
of determination and, 113–14, 117–18;
index of, 37, 45, 126, 140; interpretation
of, 45; magnitude of, 35–36; negative, 36;
positive, 36; in results section of research
reports, 35
equal expected frequencies, 176–77, 182
error: in decision making, 34; margin of, 37,
125, 201; standard of estimate, 122–23;
Type I, 34, 44, 136n, 150; Type II, 34, 44.
See also sampling error
error scores, 123, 128; reliability and, 194–95
error term, 154, 168
ethics, research, 227–28
expected frequencies, 173–82
experimental groups, 7–9, 11–13, 16, 18–19,
36, 134–35, 137–*38*, 158, 234; comparing,
12–13
experimental research, 6–18; A-B-A single-
case design in, 14; comparing groups
in, 12–13; comparing individuals
in, 13–15; control groups in, 7–9,
11–13; counterbalanced designs in, 13;
dependent variables in, 6, 8–10, 12,
14–15, 18, 32; differential selection in,
11; effect size in, 34–37; experimental
groups in, 7–8; external validity of, 9,
(threats to), 11–12; extraneous variables
in, 8–9, 13–14; Hawthorne effect in, 11;
history in, 9; independent variables in,
5–6, 9–15, 18–19; instrumentation in,
10; internal validity of, 9, (threats to),
9–11; intervention in, 5–6, 9–15, 18–19;
John Henry effect in, 11–12; matching in,
12; maturation in, 10; preexperimental
designs in, 13; quasi-experimental
designs in, 13; random assignment in, 12;
single-case (or single-subject) designs in,
13–15; statistical regression in, 10; testing

in, 10; time-series designs in, 13; true
    experimental designs in, 13
experimental studies, 7, 9, 11–13, 18, 28, 32,
    35–36, 41, 43, 201, 240; research studies,
    227, 229, 231n6, 232, 236–37
ex post facto research, 7, 15, 18, 136
external validity, 9, 14, 18; threats to, 3,
    11–12, 19
extraneous variables, 8–9, 13–14, 18–19

face validity, 205–6, 209–10
factorial analysis of variance, 151
F ratio, 137, 149, 168, 249; computing, 150–
    51, 154–57, 168, 249; interpretation of,
    157–59, 164, 167–68, 249
freedom, degrees of. *See* degrees of freedom
frequency distributions, 49–55, 50, 60, 62,
    66
frequency polygons, 53–56, 66; comparing
    histograms and, 55–56, 66
frequency tables, 52, 70, 239, 243–44
F statistic, 150, 168
F test, 150
F value, 137, 150, 168

gain scores, 7, 142
Galton, Sir Francis, 108
Gaussian Model, 90
goodness of fit, chi square test, 175–76, 182
grade equivalents (GE), 185–87, 189–91
grade point average (GPA), 23, 29, 103,
    105–6, 120, 221, 229
Graduate Record Examination (GRE), 136,
    187, 222
graphing correlation, 104–8
graphing data, 49–50, 53–67; accuracy
    in, 64–65, 67; bar graphs, 58–61, 66;
    box plots in, 62–64, 66–67; frequency
    polygons in, 53–56; histograms in, 53–56;
    line graphs, 61–62, 65–66; pie graphs,
    56–58, 66
graphing regression equations, 125–26

graphs, 66–67; bar, 48, 58–61, 66, 234;
    drawing accurate, 64–65, 67; line, 61–62,
    65–66; pie, 56–58, 66
groups, comparing, in experimental research,
    12–13
group scores, 94

Hawthorne effect, 11
heterogeneity of the group, 200, 203
histograms, 53–56, 60, 66; comparing
    frequency polygons and, 55–56, 66
homogeneity of variances, 137. *See also*
    assumption of the homogeneity of
    variances
honestly significant difference (HSD), 159
hypothesis, 4, 30–31, 43; alternative,
    31–33, 40–41, 43–44; and choosing
    the right statistical test, 218–19, 222;
    for correlation, 21, 110–11; defined,
    30; directional, 31–32, 44, 134, 138;
    nondirectional, 31–32, 44, 135; null,
    31–33, 43–44, 110–11, 117; for one-way
    analysis of variance (one-way ANOVA),
    156, 158, 168; research, 30–32; in
    research proposal, 30–32, 225, 228–32,
    236; research report and, 234–35, 237;
    statistical, 31, 40, 43–44, 112; steps in
    testing, 40–41, 46; for *t* tests, 21, 134–37;
    for two-way analysis of variance (two-way
    ANOVA), 161

incidental samples. *See* convenience samples
independent samples, *t* tests for. *See under t*
    tests: for independent samples
independent variables, 18, 224, 231n6; in
    ANOVA, 149–52, 158, 160–63, 167–69;
    chi square tests and, 173, 175, 178, 181–
    82; in experimental research, 5–6, 9–15,
    18–19; in nonexperimental research, 7,
    15; in regression, 120, 123, 126, 128
index: of effect size, 37, 45, 126, 140; of
    variability, 80, 85

individuals, comparing, in experimental research, 13–15
individual scores, 54, 94, 199
inferential statistics, 29–30, 40, 43, 72–74, 232; analysis of variance (ANOVA), 149–69; chi square test, 173–82; *t* test in, 133–47
instrumentation as threat to internal validity, 10, 19
instrument length, 200
instruments, reliability of, 201
instrument section in research proposal, 231–32, 236
intelligence tests, 10, 12, 22, 24, 84, 93, 97, 207
interaction, 161–63, 169; disordinal, 162, 169; graphing, 162–63; ordinal, 162–63, 169
interaction graph, 162, 169
intercept, 121–26, 128–29
intercorrelation tables, 115–16, 118, 245
internal consistency, measures of, 193, 195–96, 202
internal validity, 9; threats to, 3, 9–11, 19
inter-rater reliability, 193, 195, 198, 202
interval scale, 23–24, 42–43, 73, 97, 100, 175, 251
intervention in experimental research, 5–6, 9–15, 18–19
introduction: in research proposal, 225–26, 228–30, 236; in research report, 225, 233, 237
IQ tests. *See* intelligence tests

John Henry effect, 11–12
joint bars, 60–61, 66, 244

KR-20, 196–98, 202
KR-21, 196–98, 202
Kuder-Richardson reliability methods, 197–98, 202

level of significance, 103, 111, 116, 135, 137, 140, 143, 145, 156, 166, 168; probability and, 32–34, 44

limitations of study in research report, 230, 235–36
linear regression, 120, 128
linear relationships, 108–9, 117, 120, 128
line graphs, 61–62, 65–66
line of best fit, 119, 121
lines, regression, 119, 121–22, 125, 128
literature review: importance of, 5; in research proposal, 225–30, 236; in research report, 225–26, 233, 235, 237
local norms, 187, 191
longitudinal design, 16, 19–20
longitudinal studies, 17, 20
lower boundary, 40, 46
lower limit, 39–40, 46

magnitude of effect size, 35–36
main effects, 161, 164–66, 168
margin of error, 37, 125, 201
mastery, reporting scores in terms of, 190, 192
mastery learning, theory of, 190, 192
mastery tests, 85
matching groups, 12
maturation, as threat to internal validity, 10
mean, 72, 74; comparing, 36, 72–74; total, 152, 154, 166
mean, standard error of, 38–39, 45–46
mean squares (MS), 153, 156, 159, 161, 166, 168–69; between/among, 153, 168; within, 153, 161, 166, 168
measurement, 22–23, 42; standard error of, 198–99, 202–3
measurement scales, 23–24, 42–43; interval, 23–24, 42–43; nominal, 23, 42–43; ordinal, 23, 42–43; ratio, 23–24, 42–43
measures of central tendency, 69–75; comparing mode, median and mean, 72–74; defined, 69, 74; mean, 72, 74; median, 71–72, 74; mode, 70–71, 74
measures of internal consistency, 193, 195–96, 202

measures of variability, 77–87; range, 79; standard deviation and variance, 79–85; using, 83–84

median, 71–*72*, 74

methodology, 40–41; in research proposal, 225–26, 228–32, 236; in research report, 226, 233, 237

mode, 70–71, 74

multimodal distribution, 71, 74

multiple correlation, 126, 129

multiple regression, 119–20, 126–29, 218, 224

national norms, 187, 191

negative correlation, 37, 103–6, 108–9, 117

negative effect size, 36

negatively skewed distributions, 93–94, 99

nominal data, 29

nominal scale, 23, 42–43, 73–74, 174, 181, 243, 250–51

nondirectional hypothesis, 32, 44, 135, 146

nonexperimental research, 6–7, 15–18; causal comparative *(ex post facto)* research, 7, 15–16; cohort studies in, 17; cross-sectional designs in, 16–17; descriptive research, 7, 16–18; longitudinal studies in, 17; panel studies in, 17; trend studies in, 17

nonexperimental studies, 15–16

nonmastery, reporting scores in terms of, 190, 192

nonparametric statistics, 29, 43, 173; descriptive free (assumption free) statistics, 29n

normal curve, 89–100; percentile ranks and, 97–98

normal distribution, 90–91, 99

norming group, 185–91

norm-referenced tests, 16, 185–90, 210; percentile ranks, 187–88

norms, 186–91, 232; grade-equivalent, 186–87, 189–91; local, 187, 191; national, 187,

191; percentile band, 188, 191; percentile rank, 186–88, 191; stanine, 185–91

null hypothesis, 31–34, 40–41, 43–44, 110–11, 117; for ANOVA, 150, 156, 158, 160–61, 164, 167–69; for chi square test, 176–82; rejecting, 34, 41, 43–44, 111, 117, 137n4, 139–40, 143, 150, 156, 158, 160, 167, 181; retaining, 34, 41, 44, 111, 117, 140, 145, 150, 156, 176–78; for *t* test, 134n, 135, 137n4, 138–40, 143–46

numerical data, 5–6 and *passim*

observed frequencies, 173–77, 179–82

one-tailed test, 111, 117, 135–36, 140, 143, 145–46

one-way analysis of variance (one-way ANOVA), 151–60, 167, 218; to analyze survey data, 249; conceptualizing, 152–56; example of, 158–59; hypotheses for, 156, 158, 168; summary table, 156–57

ordinal data, 29

ordinal interaction, 162–*63*

ordinal scale, 23, 42–43, 73, 97, 100, 175, 251

outliers, 108, 117, 119

paired samples, *t* tests for, 40, 133, 136, 141–43, 146, 218, 223, 239, 246–49, 251

pairwise comparison, 149, 151, 168

panel studies, 17

parameters, 25, 42. *See also* population parameters

parametric statistics, 29, 43, 173–74

Pearson, Karl, 108, 174

Pearson correlation, 33, 41, 103, 218, 223–24, 239, 243; analyzing survey data, 243–46, 251; computing, 111–12, 175, 181; factors affecting, 112–13; hypotheses for, 216; interpreting, 109–10; Pearson product moment, 108–12, 117, 174, 194; requirements for, 108. *See also* correlation tables; intercorrelation tables

Pearson product moment, 108–12, 117, 174, 194

Pearson's *r*, 108–9, 117

percent correct, 186, 190–92

percentile bands, 188, 191

percentile ranks, 53, 89, 97–98, 100, 186–88, 191; grade equivalents, 189–90; normal curve and, 89, 97–98, 100; stanines, 188–*89*

percentiles, 23, 53, 97–98, 100, 185, 188–90

pie charts. *See* pie graphs

pie graphs, 56–58, 66

p level, 32

population mean, 38–40, 46, 72, 74, 138, 144, 156n, 161, 168

population parameters, 25, 30, 37, 39, 42, 45–46

populations, 24–25, 42; comparing the variance and standard distribution of, 82–83; standard deviation of, 45, 82; testing hypotheses about, 40, 46, 138; using samples to estimate values, 37–40

positive correlation, 31, 37, 104–6, 108–9, 117, 207–8

positive effect size, 36

positively skewed distributions, 93–94, 99

post hoc comparison, 156; pairwise, 149, 151, 168; Tukey method for, 159–60

practitioner research, 4–6, 17–18, 215

prediction, 119–20

predictive validity, 207–8, 210

predictor variables, 120, 126–29

preexperimental designs, 13

pretest-posttest design, 141

probability, 32–34, 44; level of significance and, 32–34, 44

procedure section in research proposal, 229, 231–32, 236–37

proportion of the variability, 123

proposal, research. *See* research proposal

pure research, 4

p value, 32–34, 41, 44–45, 103, 110–11, 113, *247,* 249; ANOVA, 159, 166–68; chi square test, 177–78, 180; *t* test, 135–36, 140, 143, 145

qualitative data, 4, 240

qualitative research, 5–6

qualitative studies, 226, 236

quantitative data, 4, 15, 226, 240

quantitative-experimental research, tools used in, 5

quantitative research, 5–6

quantitative studies, 226, 236

quasi-experimental designs, 13

random assignment, 12

random samples, 16, 26, 150, 165n6, 167

range, 86

rank-order correlation coefficient, 109

ratio scale, 23–24, 42–43, 73, 97, 100, 175, 251

raw score, 72, 80, 86, 89, 94–100, *159,* 186–87, 189–91

references: in research proposal, 226, 228–30, 232–33, 236–37; in research report, 226, 233, 237

regression, 119–29; linear, 120, 128; multiple, 119–20, 126–29, 218, 224; simple, 120–26, 128, 218

regression equations, 121–25, 128–29; graphing, 125–26

regression line, 119, 121–22, 125, 128. *See also* line of best fit

relationships: cause-and-effect, 5, 7, 15, 18; curvilinear, 108–9; linear, 108–9, 117, 120, 128; studying, 36–37

reliability, 193–203; acceptable level of, 201, 203; alternate forms, 196, 202; Cronbach's coefficient alpha, 196, 198, 202; defined, 194, 201; error score, 194–95; factors affecting, 200; of instruments, 201–2; inter-rater, 193, 195, 198, 202; Kuder-Richardson methods, 197–98, 202; measures of internal consistency, 193,

195–96, 202; methods of assessing, 195–98; split-half method, 197, 202; standard error of measurement, 198–99, 202–3; test-retest, 195–98, 202; theory of, 194–95; true score, 193–95, 199, 201–3

report, research. *See* research report

research, 3; action, 3–6, 17–18, 134; applied, 4–5, 17; basic (pure), 4, 6, 17; causal comparative, 3, 7, 15, 18, 136; descriptive, 3, 6n, 7, 16–19; ethics in, 227–28; *ex post facto,* 7, 15, 18, 136; hypothesis in, 30–32; nonexperimental, 6–7, 15–18; practitioner, 4–6, 17–18, 215; qualitative, 5–6; quantitative, 5–6; quantitative-experimental, 5; scientifically based, 12. *See also* experimental research

research hypothesis, 31, 40–41; for one-way analysis of variance (one-way ANOVA), 156; in research proposal, 225, 228–32, 236; steps in testing, 40–41; for *t* test, 134–35; for two-way analysis of variance (two-way ANOVA), 161. *See also* alternative hypothesis

research proposal, 225–33, 236–37; appendix in, 226, 232; data analysis section in, 229, 231–32, 237; instrument section in, 231–32, 236; introduction in, 225–26, 228–30, 236; literature review in, 225–30, 236; methodology in, 225–26, 228–32, 236; procedure section in, 229, 231–32, 236–37; references in, 226, 228–30, 232–33, 236–37; sample section in, 231, 236

research report, 225–27, 233–35, 237; abstract in, 226, 233, 237; appendix in, 226, 233–34; discussion in, 9, 226–27, 230, 233, 235, 237; introduction in, 225, 233, 237; limitations of the study in, 230, 235–36; literature review in, 225–26, 233, 235, 237; methodology in, 226, 233, 237; references in, 226, 233, 237; results in, 35, 225–26, 233–35, 237

research studies, planning and conducting, 225–37

residual, 122, 128, 164, 166n

results section in research report, 35, 225–26, 233–35, 237

robust statistic, 137, 146, 150, 167

sample bias, 28–29, 43

samples, 24–25, 42; comparing variance and standard distribution of, 82–83; computing variance and SD for, 82–83; convenience, 27–28; in estimating population values, 37–40; incidental, 27–28; mean of, 38–39; random, 16, 26, 150, 165n6, 167; simple random, 26; stratified, 27; systematic, 26–27

sample section in research proposal, 231, 236

sample size, 12, 18, 26, 28–29, 33–35, 38–39, 42–43, 83, 86, *91,* 158, 235; chi square test, 175, 181; correlation, 110, 116; survey data, statistical tests and, 243, 246n, 247; *t* test, 134, 137–39, 141

sample statistics, 25, 30, 37, 45

sampling, methods of, 25–28

sampling error, 25, 32–33, 37–39, 42, 45, 134–35, 146, 158, 179

sampling statistics, 29, 43

scale score, 95, 99, 185–86, 191

scattergrams, 103–9, 112–13, 117, 119, *121, 125*

scatterplots. *See* scattergrams

scenarios, choosing statistical tests for, 215–24

Scholastic Aptitude Test (SAT), 120, 185, 187, 207

scientifically based research, 12

scores: deviation, 79–81, 86; group, 94; individual, 54, 94, 199; scale, 95, 99, 185–86, 191; standard, 89, 94–99; true, 193–95, 199, 201–3. *See also* raw score; *z* scores

scores for standardized tests, interpreting, 185–92

shared variance, 113, 117, 123n3

significance, probability and level of, 32–34

simple random sample, 26, 42

simple regression, 120–26, 128, 218; example of, 123–25

single-case (or single-subject) designs, 13–14; problems associated with, 14–15; quantitative data in, 15. *See also* A-B-A single-case design

single sample, *t* tests for, 136, 143–46, 218

single-subject design, 13, 19

skewed distributions, 89, 93–94, 99

slope, 121–*22*, 124–26, 128–29

Spearman-Brown prophecy formula, 197, 200

Spearman rank-order correlation, 109

split-half method, 197, 202

SPSS, 33–34, 40, 46, 52, 70, 95n4, 98, 111n4, 137, 180

squared correlation coefficients, 36–37

squared units, 83, 86

square root of the variance, 81–83, 86

squaring the standard deviation, 80, 82, 85

standard deviation (SD), 36, 38–40, 45–46, 80, 86–87; computing for populations and samples, 82–83; in distributions with extreme scores, 84–85; in effect size, 35; factors affecting, 85; of a population, 45, 82; squaring, 80, 82, 85; using, 83–84; variance and, 79–85

standard error of estimate (SE), 119, 122–23, 125, 128

standard error of measurement (SEM), 198–99, 202–3

standard error of the mean, 38–39, 45–46

standardized test scores, 185–92

standard scores, 89, 94–99; normal curve and percentile rank, 97–98; other converted scores, 97; *T* scores, 96, 100; *z* scores, 95–96, 99

Stanford-Binet IQ test, 97

stanines, 185–91

statistic: robust, 137, 146, 150, 167; versus statistics, 25n, 42

statistical analysis, 41

statistical hypothesis, 31, 40, 43–44, 112. *See* null hypothesis

statistical regression, as threat to internal validity, 10

statistical significance, 32–33, 44–45

statistical symbols, list of, vii–viii

statistical tests: choosing the right, 215–24; using to analyze survey data, 239–51

statistics, 21–46; assumption-free, 43; defined, 22; distribution-free, 29n, 181; nonparametric, 29, 43, 173; parametric, 29, 43, 173–74; sample, 25, 30, 37, 45; sampling, 29, 43; versus *a statistic,* 25n, 42. *See also* descriptive statistics; inferential statistics

stratification, 186

stratified sample, 27, 42

sum of squares (SS), 152, 153, 156, 161, 166, 168; between groups, 152, 168; within groups, 152, 168; total, 152, 168

surveys: analyzing survey data, 239–51; constructing, 241–43; definition, 240–41

systematic sample, 26–27, 42

table of critical values, 33–34, 111, 135, 139–40, 143, 146, 176

tables: contingency, 179; correlation, 103, 116, 118; intercorrelation, 115–16, 118, 245

t distribution, 134–35, 143

test bias, 205, 209–10

testing, as threat to internal validity, 10

testing, hypothesis, 40–41, 46

test of best fit, 173, 218

test of independence, 173, 175, 178–79, 182, 218, *220,* 223–24, 250

test-retest reliability, 195–98, 202

tests: commercial achievement, 206; content referenced, 190n; criterion-referenced,

85, 185–86, 190, 192; domain-referenced, 190n; interpreting scores on, 185–92; norm-referenced, 16, 185–90, 210; standardized, 185–92; statistical, choosing the right, 215–24; statistical, using to analyze survey data, 239–51. *See also* intelligence tests

time-series design, 13

total mean, 152, 154, 166

total sum of squares (SST), 152, 168

trend studies, 17

true experimental design, 13

true score, 193–95, 199, 201–3

T scores, 96–*97*, 100

t tests, 29, 133–47

to analyze survey data, 246–*49*; hypotheses for, 21, 134–37; for independent samples, 32, 35, 136–41, 146, 149, 150n1, 167, 216, 218–19, 234, 239, 246, 251; for paired samples, 40, 133, 136, 141–43, 146, 218, 223, 239, 246–49, 251; for single sample, 136, 143–46, 218; using, 246–*49*

t-test table of critical values, 139–40, 143, 146

Tukey, John, 62

Tukey method of post hoc multiple comparisons, 159–60, 168

two-tailed test, 111, 117, 135–36, 145–46

two-way analysis of variance (two-way ANOVA), 160–69, 218; conceptualizing, 160–61; example of, 165–67; graphing the interaction, 162–64; hypotheses for, 161

two-way ANOVA summary table, 164–66

Type I error, 34, 44, 136n, 150

Type II error, 34, 44

unequal expected frequencies, 177–*78*, 182

upper boundary, 40, 46

upper limit, 39–40, 46

validity, 205–11, assessing, 104, 205, 209–10; concurrent, 207, 210; construct, 208, 210; content, 206–7, 210; criterion-related, 207–8, 210; defined, 206, 210; external, 9, 14, 18, (threats to), 3, 11–12, 19; face, 209–10; internal, 9, (threats to), 3, 9–11; predictive, 208–8, 210

validity coefficient, 207–10

variability: index of, 78, 85; measures of, 77–87; proportion of, 123

variables, 22–24, 42; continuous, 22; criterion, 122–23, 128; discrete, 22; extraneous, 8–9, 13–14, 18–19. *See also* dependent variables; independent variables

variance, 86–87; computing, 81–83; defined, 80; in distributions with extreme scores, 84–85; factors affecting, 85; homogeneity of, 137; shared, 113, 117, 123n3; using, 83–84

variation: between groups, 161, 169; within-groups, 161, 169

Wechsler IQ test, 97

within-groups mean square (SS_w), 154

within-groups sum of squares (SS_w), 152, 168

within-groups variation, 161, 169

z scores, 40, 73, 89, 95–100, 191; negative, 96, 100; positive, 96, 100

About the Author

Ruth Ravid is professor emerita of education at National Louis University. She holds an MA and a PhD in education from Northwestern University. Ruth has coauthored several books; *Action Research in Education: A Practical Guide* (with S. E. Efron, 2013); A *Study Guide for Practical Statistics for Educators* (with E. Oyer, 2011); and *Practical Statistics for Business: An Introduction to Business Statistics* (with P. Haan, 2008). Ravid is the coeditor of *Collaboration in Education* (with J. J. Slater, 2010) and *The Many Faces of School-University Collaboration: Characteristics of Successful Partnerships* (with M. G. Handler, 2001); as well as a coeditor of the journal *i.e.: Inquiry in Education.*